Lecture Notes in Computer Science 13959

The series Lecture Notes in Computer Science (LNCS), including its subseries Lecture Notes in Artificial Intelligence (LNAI) and Lecture Notes in Bioinformatics (LNBI), has established itself as a medium for the publication of new developments in computer science and information technology research, teaching, and education.

LNCS enjoys close cooperation with the computer science R & D community, the series counts many renowned academics among its volume editors and paper authors, and collaborates with prestigious societies. Its mission is to serve this international community by providing an invaluable service, mainly focused on the publication of conference and workshop proceedings and postproceedings. LNCS commenced publication in 1973.

Daniel Gruss · Federico Maggi ·
Mathias Fischer · Michele Carminati
Editors

Detection of Intrusions and Malware, and Vulnerability Assessment

20th International Conference, DIMVA 2023
Hamburg, Germany, July 12–14, 2023
Proceedings

 Springer

Editors
Daniel Gruss 📵
Graz University of Technology
Graz, Austria

Federico Maggi 📵
AWS Italy
Milan, Italy

Mathias Fischer 📵
Universität Hamburg
Hamburg, Germany

Michele Carminati 📵
Politecnico di Milano
Milan, Italy

ISSN 0302-9743   ISSN 1611-3349 (electronic)
Lecture Notes in Computer Science
ISBN 978-3-031-35503-5   ISBN 978-3-031-35504-2 (eBook)
https://doi.org/10.1007/978-3-031-35504-2

This Springer imprint is published by the registered company Springer Nature Switzerland AG
The registered company address is: Gewerbestrasse 11, 6330 Cham, Switzerland

Preface

On behalf of the Program Committee, we are delighted to present the proceedings of the 20th Conference on Detection of Intrusions and Malware & Vulnerability Assessment, also known as DIMVA. It is an honor to be part of the team organizing the conference's 20th anniversary. Over the past two decades, DIMVA has become a well-established security conference, consistently attracting high-quality research and fostering collaboration between academia, industry, and government. DIMVA is organized by the Special Interest Group – Security, Intrusion Detection, and Response (SIDAR) - of the German Informatics Society (GI).

This year, we received 43 valid submissions and accepted 13 papers (12 full papers and 1 short paper), resulting in a competitive acceptance rate of 30.2%. All papers received three double-blind reviews, and each PC member was assigned up to three papers to review.

For the first time, DIMVA introduced multiple submission deadlines, with a balanced acceptance rate across both cycles: 6 out of 22 papers were accepted in the first cycle, and 7 out of 21 papers were accepted in the second cycle. The number of submissions increased slightly, and in response, we implemented a revision model which closely follows the concept of the previous "accept with shepherding" decision but allows for more significant changes in the revised papers.

We would like to express our deep gratitude to the Program Committee members for their countless hours spent reviewing submissions for both cycles, engaging in online and in-person discussions, and shepherding numerous papers. Online alone, PC members and external reviewers exchanged more than 400 comments. The in-person PC meeting in Graz helped with discussing the remaining undecided papers and converging to a final selection of papers and revision criteria.

We also would like to thank the Organizing Committee for their dedication and hard work, laying the foundation for DIMVA's successful 20th anniversary. We appreciate the support from our sponsors and host institutions.

Finally, we would like to thank the authors who submitted their work and presented their contributions, as well as the attendees who participated in the conference. Your enthusiasm and commitment have made DIMVA the thriving and enriching event it is today. We look forward to your future contributions.

May 2023

Daniel Gruss
Federico Maggi

Organization

General Chair

Mathias Fischer   Universität Hamburg, Germany

Program Committee Chairs

Daniel Gruss   Technische Universität Graz, Austria
Federico Maggi   Amazon Web Services, Italy

Steering Committee Chairs

Ulrich Flegel   Infineon Technologies, Germany
Michael Meier   University of Bonn and Fraunhofer FKIE, Germany

Steering Committee

Magnus Almgren	Chalmers University of Technology, Sweden
Sébastien Bardin	CEA, France
Gregory Blanc	Télécom SudParis, France
Herbert Bos	Vrije Universiteit Amsterdam, The Netherlands
Danilo M. Bruschi	Università degli Studi di Milano, Italy
Roland Bueschkes	RWE AG, Germany
Juan Caballero	IMDEA Software Institute, Spain
Lorenzo Cavallaro	King's College London, UK
Hervé Debar	Télécom SudParis, France
Sven Dietrich	City University of New York, USA
Cristiano Giuffrida	Vrije Universiteit Amsterdam, The Netherlands
Bernhard Haemmerli	Acris GmbH and HSLU Lucerne, Switzerland
Thorsten Holz	CISPA Helmholtz Center for Information Security, Germany
Marko Jahnke	CSIRT, German Federal Authority, Germany
Klaus Julisch	Deloitte, Switzerland
Christian Kreibich	ICSI, USA

Christopher Kruegel	UC Santa Barbara, USA
Pavel Laskov	University of Liechtenstein, Liechtenstein
Federico Maggi	Amazon Web Services, Italy
Clémentine Maurice	CNRS, IRISA, France
Roberto Perdisci	University of Georgia and Georgia Institute of Technology, USA
Michalis Polychronakis	Stony Brook University, USA
Konrad Rieck	TU Braunschweig, Germany
Jean-Pierre Seifert	Technical University Berlin, Germany
Robin Sommer	ICSI/LBNL, USA
Urko Zurutuza	Mondragon University, Spain

Program Committee

Magnus Almgren	Chalmers University of Technology, Sweden
Daniel Arp	University College London, UK
Sébastien Bardin	CEA, France
Leyla Bilge	Norton Research Group, France
Gregory Blanc	Télécom SudParis, Institut Polytechnique de Paris, France
Jorge Blasco	Universidad Politécnica de Madrid, Spain
Kevin Borgolte	Ruhr-University Bochum, Germany
Juan Caballero	IMDEA Software Institute, Spain
Michele Carminati	Politecnico di Milano, Italy
Daniele Cono D'Elia	Sapienza University of Rome, Italy
Andrea Continella	University of Twente, The Netherlands
Marco Cova	VMware, UK
Lesly-Ann Daniel	KU Leuven, Belgium
Herve Debar	Télécom SudParis, France
Sven Dietrich	City University of New York, USA
Bernhard Haemmerli	Acris GmbH and HSLU Lucerne, Switzerland
Christophe Hauser	University of Southern California, USA
Johannes Kinder	Bundeswehr University Munich, Germany
Andreas Kogler	Graz University of Technology, Austria
Christopher Kruegel	University of California, Santa Barbara, USA
Pavel Laskov	University of Liechtenstein, Liechtenstein
Moritz Lipp	Amazon Web Services, Austria
Michael Meier	University of Bonn and Fraunhofer FKIE, Germany
Nick Nikiforakis	Stony Brook University, USA
Anita Nikolich	University of Illinois at Urbana Champaign, USA

Fabio Pagani	UC Santa Barbara, USA
Tapti Palit	Purdue University, USA
Feargus Pendlebury	Meta, UK
Fabio Pierazzi	King's College London, UK
Mario Polino	Politecnico di Milano, Italy
Michalis Polychronakis	Stony Brook University, USA
Kaveh Razavi	ETH Zurich, Switzerland
Konrad Rieck	TU Berlin, Germany
Michael Schwarz	CISPA Helmholtz Center for Information Security, Germany
Martin Schwarzl	Graz University of Technology, Austria
Lea Schönherr	CISPA Helmholtz Center for Information Security, Germany
Jean-Pierre Seifert	Technical University of Berlin, Germany
R. Sekar	Stony Brook University, USA
Daniele Sgandurra	AI4Sec, Huawei Technologies, Germany
Deborah Shands	SRI International, USA
Seungwon Shin	KAIST, South Korea
Sandra Siby	EPFL, Switzerland
Guillermo Suarez-Tangil	IMDEA Networks Institute, Spain
Kimberly Tam	University of Plymouth, UK
Juan Tapiador	Universidad Carlos III de Madrid, Spain
Flavio Toffalini	EPFL, Switzerland
Jo Van Bulck	KU Leuven, Belgium
Erik van der Kouwe	Vrije Universiteit Amsterdam, The Netherlands
Jan Wichelmann	University of Lübeck, Germany
Christian Wressnegger	Karlsruhe Institute of Technology (KIT), Germany
Roland Yap	National University of Singapore, Singapore
Stefano Zanero	Politecnico di Milano, Italy
Urko Zurutuza	Mondragon Unibertsitatea, Spain

Additional Reviewer

Flavien Solt	ETH Zurich, Switzerland

Contents

Side Channels Attacks

MAMBO–V: Dynamic Side-Channel Leakage Analysis on RISC–V

Jan Wichelmann$^{(\boxtimes)}$, Christopher Peredy, Florian Sieck, Anna Pätschke, and Thomas Eisenbarth

University of Lübeck, Lübeck, Germany
{j.wichelmann,c.peredy,florian.sieck,a.paetschke,
thomas.eisenbarth}@uni-luebeck.de

Abstract. RISC–V is an emerging technology, with applications ranging from embedded devices to high-performance servers. Therefore, more and more security-critical workloads will be conducted with code that is compiled for RISC–V. Well-known microarchitectural side-channel attacks against established platforms like x86 apply to RISC–V CPUs as well. As RISC–V does not mandate any hardware-based side-channel countermeasures, a piece of code compiled for a generic RISC–V CPU in a cloud server cannot make safe assumptions about the microarchitecture on which it is running. Existing tools for aiding software-level precautions by checking side-channel vulnerabilities on source code or x86 binaries are not compatible with RISC–V machine code.

In this work, we study the requirements and goals of architecture-specific leakage analysis for RISC–V and illustrate how to achieve these goals with the help of fast and precise dynamic binary analysis. We implement all necessary building blocks for finding side-channel leakages on RISC–V, while relying on existing mature solutions when possible. Our leakage analysis builds upon the modular side-channel analysis framework Microwalk, that examines execution traces for leakage through secret-dependent memory accesses or branches. To provide suitable traces, we port the ARM dynamic binary instrumentation tool MAMBO to RISC–V. Our port named MAMBO–V can instrument arbitrary binaries which use the 64-bit general purpose instruction set. We evaluate our toolchain on several cryptographic libraries with RISC–V support and identify multiple leakages.

Keywords: RISC-V · Side-channel attacks · Dynamic binary instrumentation · Software security

1 Introduction

Executing workloads in cloud environments with shared hardware resources is becoming more and more important, promising great flexibility and scalability. From a security viewpoint, however, this trend comes with a number of challenges, as shown by manifold examples of attacks that exploit microarchitectural side-channels in cloud systems [21, 22, 53].

© The Author(s), under exclusive license to Springer Nature Switzerland AG 2023
D. Gruss et al. (Eds.): DIMVA 2023, LNCS 13959, pp. 3–23, 2023.
https://doi.org/10.1007/978-3-031-35504-2_1

While most of these cloud systems and the corresponding attacks are based on the conventional x86 architecture, a new architecture called RISC–V is gaining traction in both embedded applications and general-purpose hardware. The royalty-free license [4] of RISC–V enables affordable hardware through lower development costs, and helps innovation: For example, there now are several open-source CPU designs which can be analyzed and extended by anyone [26,33,45], promising the development of new hardware features like secure trusted execution environments (TEEs) which avoid the issues of existing commercial solutions. The software support for the RISC–V platform is growing as well, with major compiler vendors adding backends for emitting RISC–V machine code, which in turn allows porting operating systems like Linux.

The growing importance of RISC–V in general-purpose and cloud computing, coupled with a wide spectrum of CPU designs from various vendors, still necessitates caution to prevent repeating the mistakes that caused a lot of security issues on the established platforms. One particular example is *microarchitectural timing leakage* in cryptographic libraries, where subtle differences in how the microarchitecture processes certain operations lead to exploitable leakages, allowing a co-located attacker running on the same hardware as the victim code to extract cryptographic secrets. By microarchitectural timing leakage, we refer to architectural traces only, excluding transient execution attacks. As most of the existing RISC–V hardware finds usage in the IoT or the automotive domain, there has been more focus on physical attacks like power side-channels, and little work on analyzing the co-location scenario so far. However, it is likely that many attack vectors from x86 and ARM will apply to RISC–V systems as well. While there are several proposals for hardware countermeasures that would address this issue (e.g., resistant cache designs [11,43,49]), it is unlikely that all CPU vendors will include one of those mitigations in their processors. Thus, absent a proven hardware-based countermeasure, software-level mitigations are needed.

By now, most established libraries address timing leakages by employing so-called *constant-time code*, i.e., code that exhibits the same control flow and memory access pattern independent of its secret inputs. However, the new compiler backends and different instruction set of RISC–V may re-introduce leakage previously fixed at source level [3,10]. In addition, there is ongoing work on assembly-level implementations of cryptographic primitives, which are carefully optimized to fully utilize the underlying hardware to achieve best performance [44], but may have subtle leakages. While there are lots of approaches for finding leakages on source-level or via generic languages, those cannot detect leakage introduced by the compiler. Finally, most of the corresponding proof-of-concept implementations lack usability [23] or do not apply to RISC–V.

In this work, we discuss the requirements of analyzing RISC–V software for side-channel leakages, and show how an established side-channel analysis framework can be adapted to also support RISC–V binaries. For that, we build upon the *Microwalk* framework [51], that analyzes execution traces in order to identify vulnerabilities, and then yields a detailed leakage report. While Microwalk generates its execution traces through dynamic binary instrumentation (DBI), no such tool is yet available for RISC–V. Thus, we develop the first DBI tool for RISC–V, called MAMBO–V, which sets up on the MAMBO toolkit [18] for ARM, and show how we can use this tool to generate Microwalk-compatible traces. We evaluate our leakage analysis toolchain on several cryptographic libraries with support for RISC–V, and uncover multiple vulnerabilities.

1.1 Our Contribution

In summary, our contributions are:

- We analyze the similarities and differences between RISC–V and established architectures in terms of side-channel vulnerabilities, and extract requirements for building side-channel-resistant software on RISC–V.
- We implement MAMBO–V, a RISC–V port of the ARM-based DBI tool MAMBO, enabling us to natively instrument RISC–V binaries.
- We include MAMBO–V in the Microwalk framework for finding timing side-channels in software binaries, building the first toolchain for automatically analyzing RISC–V programs.
- We analyze several RISC–V builds of cryptographic libraries and detect various leakages.

The source code is available at https://github.com/UzL-ITS/MAMBO-V.

Responsible Disclosure. We disclosed the potentially exploitable AES vulnerabilities to the developers of the respective libraries, who all acknowledged our findings. They were mostly aware of the issues of the relevant implementations, and WolfSSL and OpenSSL have (undocumented) compiler flags which partially fix the leakages (see Sect. 6.3). At the time of submission, there is ongoing work on patches that ensure that the default implementations are secure, or on appropriate documentation changes.

2 Background

2.1 RISC–V

RISC–V is a reduced instruction set computer (RISC) load-store architecture, with a focus on broad availability through permissive licensing and high modularity to support all applications from small low-power IoT devices over personal mobile devices to large-scale general purpose computers. Its open-source character allows easy extensibility through a so-called base-plus-extension instruction set architecture (ISA). As a RISC architecture, only designated instructions operate on memory, whereas the arithmetic merely happens in registers. The most important standardized extensions for RISC–V are I, M, A, C, F, D, Zicsr and Zifencei, which are often grouped together as *RV64GC*. Also, more specialized extensions are drafted and partially ratified, such as the vector extension and scalar cryptographic extension [41]. Instruction encodings are designed to simplify hardware implementations to increase performance and efficiency [47].

2.2   Dynamic Binary Instrumentation

Binary instrumentation allows inserting code into an existing binary in order to monitor or modify the program's behavior. The insertion points are determined through user-supplied rules or callback functions.

Static binary instrumentation (SBI), also called binary rewriting, permanently inserts instrumentation code into the binary in an offline phase [12]. While this approach promises a small runtime overhead, it is error-prone due to relying on a correct disassembly of the program. In addition, SBI cannot handle special cases like just-in-time compilation or self-modifying code.

In *dynamic binary instrumentation* (DBI), the instrumentation code is added with the help of an instrumentation framework at runtime. The DBI framework combines application and instrumentation code and executes the resulting code directly on the target platform. DBI engines introduce a slightly higher overhead than SBI due to the code translation at runtime, but most prevalent instrumentation frameworks feature optimizations like caching, so each code block needs to be instrumented only once. Popular DBI engines include Intel Pin [30], DynamoRIO [9], QBDI [39] and the heavyweight analysis framework Valgrind [35], which were initially built for x86 and then, in some cases, extended to also support other architectures like ARM's AArch32 and AArch64.

However, as ARM is a RISC architecture and thus quite different to x86, x86-specific optimizations in a DBI engine may have little or even negative effects. MAMBO [18] is a DBI tool specifically designed and developed for ARM, making it suitable for efficiently handling RISC architectures. In addition to some ARM-specific optimizations, MAMBO has general DBI features like a cache for storing already instrumented code and scanning new code in basic block units. Moreover, it supports behavioral transparency, which means that the execution of all ABI-compliant binaries is guaranteed to be correct. The application binary interface (ABI) defines the calling convention, which includes register allocation for parameters and stack pointer behavior.

2.3   Microarchitectural Side-Channels

In a cloud setting, usually, many processes from different customers share the same underlying hardware. These processes may work with sensitive data, which should not be leaked to an attacker. While there are many architectural safeguards in place to prevent data from flowing from one process to another directly, there are more subtle *side-channels* that use properties of the underlying microarchitecture to extract some information from the running code. One prominent example are so-called *cache attacks* [1,7,37,53], where the attacker brings the (shared) CPU cache into a known state, and then monitors changes to this state in order to learn whether the victim has accessed data within a certain address range. This way, the attacker can infer the code line the victim is currently executing, or determine the index of a table lookup. Besides the cache, there are many more shared resources that the attacker can monitor and exploit, like the translation look-aside buffer [19] and the branch prediction unit [2]. Note that we only consider attacks that target architectural traces, so transient execution attacks like Spectre [25] are out-of-scope.

A commonly used software-based countermeasure against side-channel attacks is constant-time code without any secret-dependent memory accesses or branches [3]. This code exhibits the same control flow and data flow independent of the processed secret, so a side-channel attacker cannot learn anything by looking at an execution trace as provided by a cache attack. As cryptographic implementations are a primary target for side-channel attacks, most current cryptographic libraries feature constant-time code.

Leakage Detection Tools. To ease checking implementations for side-channel vulnerabilities, numerous tools and approaches have been proposed. Tools that analyze source code include *ct-fuzz* [20] that uses a specialized form of fuzzing, *ct-verif* [3] based on formal verification methods and *CaSym* [8] that symbolically executes the source code. Moreover, there are various tools that analyze binaries through static techniques, like *BINSEC/REL* [10] using symbolic execution, *CacheS* [46] combining symbolic execution with taint analysis, or *CacheAudit* [14] which uses formal methods to find leakages on all paths of a program. Finally, dynamic binary approaches comprise statistical timing measurements like in *dudect* [40], constraint modeling in *Abacus* [5], as well as trace alignment in *DATA* [48] or trace merging in Microwalk [51].

3 Overview

We first describe requirements and our approach for analyzing the side-channel security of RISC–V implementations running in a co-located setting.

3.1 Analysis Approach

As described in Sect. 2.3, there are numerous tools and approaches for finding side-channel leakages in software. Any useful tool should unify the following properties [23,51]: First, it should accurately localize the respective leakages, so the developer can directly understand the cause of a leakage and start building a patch. Then, the analysis should be fast enough, so there is immediate feedback whenever there is a code change. Finally, to aid adoption in the developer community, the tool should not be too hard to set up and use.

To check whether RISC–V code is leakage-free, focusing on the source code alone is insufficient. For example, there have been cases where a misguided compiler pass "optimized" constant-time code, producing binaries with leakages that are not present in the source code [3,24]. Daniel et al. [10] further provide an extensive evaluation of different compiler versions, optimization levels and target architectures, showing that constant-time properties always need to be validated on the binary level. Compiling the code for x86 and using existing analysis tools is not sufficient either, as x86 compilers may use different optimization passes than RISC–V compilers. In addition, x86 has special extensions like AES-NI or the `pclmulqdq` instruction for carry-less multiplication (used in Galois counter mode), which may substitute otherwise leaking code paths.

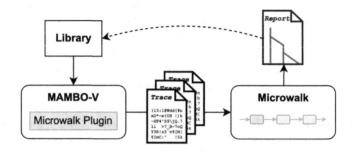

Fig. 1. RISC–V side-channel analysis overview. MAMBO–V instruments a RISC–V library and generates execution traces, which are subsequently analyzed using Microwalk. The resulting analysis report then helps the developer to find and fix the identified leakages.

The necessity to work with RISC–V specific assembly leaves the option to use either static or dynamic binary analysis. While static binary approaches offer some guarantees that purely dynamic tools cannot give, they often suffer from poor performance and require lots of manual interaction. On the other hand, dynamic analysis is heavily dependent on the achieved coverage, i.e., leakage can only be found in code that is actually executed. However, for cryptographic implementations, it was found that a small number of random test cases is sufficient to cover the relevant code [48,51]. In addition, dynamic analysis is easy to use, as the user only has to call the respective primitives.

3.2   Toolchain

With the aforementioned requirements in mind, we picked the Microwalk framework [50,51] as a basis for our RISC–V leakage analysis. Microwalk uses DBI to generate execution traces from user-supplied programs, and offers several analysis modules that compare these traces in order to find leakage. While the authors originally designed Microwalk for x86 binaries, its modular structure and generic trace format encourage addition of trace generators for other architectures.

This leaves the problem of generating Microwalk-compatible execution traces for RISC–V. At the time of writing, there is no generic DBI framework for RISC–V available, that offers the necessary flexibility for generating the information Microwalk needs. Another requirement is transparency, such that the execution traces are not influenced by the DBI engine itself, which would otherwise distort the analysis result. Instead of building a new DBI framework, we decided to port an existing framework for another RISC architecture, that is MAMBO [18] for ARM. The similarities between ARM and RISC–V allow us to reuse most of the general-purpose logic from MAMBO, like plugin handling or memory management. Our port, named MAMBO–V, implements the most significant performance optimizations from MAMBO, which are inline hash table lookups and direct branch linking. Additionally, we add support for atomic sequences, which need special handling on RISC–V hardware. We are working with the maintainers of MAMBO to contribute our RISC–V patches to the main project.

The resulting toolchain is illustrated in Fig. 1.

4   MAMBO–V Implementation

We now describe our RISC–V port of the MAMBO DBI framework, named MAMBO–V. We give an overview over its generic features and discuss notable performance optimizations as well as RISC–V specifics to be considered.

4.1   Instrumentation Approach

Target Platform. MAMBO–V targets RV64GC platforms, i.e., processors with support for the RV64I base instruction set and its most common extensions. Like MAMBO, MAMBO–V aims for *behavioral transparency*: Binaries that are compliant to the standard RISC–V ABI are executed correctly. This does not affect the correctness of our side-channel analysis, as we can expect that compilers emit standard-compliant code and that the analyzed programs are not malicious.

Execution Model. Just as the ARM implementation of MAMBO, MAMBO–V unifies the instrumentation framework and the target application in a single process. On startup, a custom ELF loader reads the RISC–V ELF file and potential dependencies of the target application into the memory of the MAMBO–V process, such that the engine can access the target's full code. After initialization is done, MAMBO–V's dispatcher proceeds loading and translating chunks of the target's code on-the-fly, while inserting instrumentation at the points specified by the user. Each chunk consists of a single basic block, i.e., a sequence of instructions with a single entry point at the beginning and a single exit point at the end. This way, the dispatcher can safely hand over control to the translated chunk, and reclaim it after the chunk has fully executed.

Plugin API. In order to facilitate the usage of MAMBO–V for application developers who want to analyze their applications, we also ported the plugin API from MAMBO. A plugin contains user-supplied functions, which are called at certain events, e.g., when translating a basic block. With these functions, the user can then insert instrumentation code during translation. Other supported events are function entry/exit, threads and system calls. In our analysis, we primarily utilize the instrumentation to insert trace writing code.

Optimizations. To speed up analysis, we have ported a number of performance optimizations from MAMBO. Most of the overhead that arises during DBI comes from the code translation and context switches between the dispatcher and the target application. The most notable optimization is the code cache, which is a common feature of DBI frameworks: It is located outside the target application's address space and stores a limited amount of translated basic blocks. This avoids re-translation of frequently executed code, improving overall performance significantly. Other optimizations are hash tables for faster resolution of translated blocks and direct branch linking to speed up jumping between different blocks in the code cache without invoking a costly context switch to the dispatcher.

4.2　New Features for RISC–V

Atomic Sequences. A challenge we encountered on RISC–V cores are tightly constrained *atomic sequences*, which ensure exclusive memory operations for multiprocessor systems and process synchronization. Software locks for resources that should only be accessed by a single thread or process at a time are often translated to *atomic loops* by the compiler. An atomic loop contains an atomic sequence, which begins with a load-reserved (LR) instruction and ends with a store-conditional (SC) instruction. The atomic loop loops over the atomic sequence until the SC eventually succeeds. The result of the SC instruction depends on whether the reserved value was accessed during the atomic sequence and on the environmental constraints defined by the ISA. Among others, the ISA defines a maximum of 16 consecutive instructions between LR and SC, and allows only the base (I) instruction set, disallowing loads, stores, backward jumps or calls.

While the compiler enforces the constraints within an atomic sequence, the instrumentation done by MAMBO–V can insert arbitrary instructions that break one of the above constraints. Figure 2 shows an example of how a direct port of MAMBO would add unconstrained instructions to an atomic sequence: First, the original loop in Fig. 2a is split into two blocks because of the conditional branch in line 3. Then, the resulting code cache blocks undergo optimization and are instrumented as shown in Fig. 2b, leading to the insertion of unconstrained instructions (line 3–5). The result is a non-sequential sequence that includes loads, stores, calls, and potential backward jumps, and is therefore not guaranteed to succeed on RISC-V. However, requiring all instrumentation to adhere to the constraints would cause some instrumentation features to be lost in the process.

On ARM, where atomic sequences are available as well, MAMBO allows users to freely insert instrumentation, which when breaking a constraint causes undefined behavior, but does not affect stability on ARM Cortex processors. However, on our SiFive U54 core, violating a constraint can block the SC instruction from succeeding entirely, leaving the process stuck in a deadlock. We encountered such a deadlock when instrumenting the dynamic linker.

Thus, for reliable instrumentation on RISC–V cores, we designed a lightweight and behaviorally transparent solution for handling atomic sequences:

```
# a0: value to store          1:block1:
# a1: lock status             2:    LR.D a1, (s3)
# s3: memory address          3:    <branch condition evaluation>
                              4:    <call trace_conditional_branch>
1:loop:                       5:    <cond. branch to block1 or block2>
2:    LR.D a1, (s3)
3:    BNE a1, zero, loop      6:block2:
4:    SC.D a1,    a0, (s3)    7:    SC.D a1,    a0, (s3)
5:    BNE a1, zero, loop      8:    BNE a1, zero, loop
```

(a) Original lock-acquire-loop.   (b) Instrumented lock-acquire-loop.

Fig. 2. Exemplary instrumentation of a lock-acquire-loop: The instrumentation may insert unconstrained instructions (marked in **blue**) into the atomic sequence, e.g., add a function call with parameters to trace a conditional branch instruction. In order to set the argument registers, the original register contents have to be written to the stack using an unconstrained store instruction. (Color figure online)

We use hardware-assisted software emulation to relax the hardware constraints by replacing the LR and the SC instructions. The LR is replaced by an equivalent normal load instruction, which marks the beginning of the software-emulated atomic sequence. To emulate the reserve, we also back up the original value for later comparison. The subsequent code is not bound by constraints anymore and safe for arbitrary instrumentation. Finally, we replace the SC instruction with a semantically equivalent atomic sequence that conditionally stores the new value if the value at the destination is equal to the previously created backup. Since we include a native atomic sequence to check for changes at the destination, our emulation remains thread-safe. The observable behavior of the emulated atomic sequence is nearly identical to the original, with the only difference being that the emulation cannot detect stores on the reserved value that do not modify it. To the best of our knowledge, this difference does not effectively change the semantics of the emulated sequence, and therefore the traces remain identical.

Global Pointer and Thread Pointer Register. In contrast to ARM, the RISC–V standard calling convention defines a global pointer register gp and a thread pointer register tp. Applications use these registers to access structures such as the global offset table and global/thread-local variables. MAMBO–V does not share these structures with its client, so gp and tp must be updated on each of the context switch between MAMBO–V and the client. Originally, on ARM, a unidirectional context switch was sufficient, as the dispatcher does not make assumptions on register contents on entry. Thus, only the context of the client is fully saved when entering the MAMBO–V context and restored when leaving again. To support the distinct gp/tp contexts on RISC–V, we implemented a full context switch for these two registers, while keeping the unidirectional context for all other registers to minimize the overhead.

Shorter Jump Encoding. RISC–V and ARM do not have direct branch instructions that take an absolute immediate address. Due to different instruction encodings, the maximum range of ARM branch instructions is ±128 MiB, while on RISC–V it is only ± 1 MiB. The code cache in MAMBO–V can be much larger than 1 MiB. Hence, for MAMBO–V, we decided to use indirect jumps to transfer control flow back to the dispatcher. Loading the address and performing the jump takes 14 additional bytes in the code cache, but due to the long lifetime of translated code and runtime overhead of the client-dispatcher context switch the effect on the overall performance and memory consumption is negligible.

5   Side-Channel Leakage Analysis.

In the following, we describe our approach for finding architecture-specific leakage in code compiled for RISC–V with the help of MAMBO–V. We focus on implementations of cryptographic algorithms, as their impact on the security of systems and communication is high. However, the concepts do apply to any scenario where secret information should not be exposed to an attacker recording execution traces. As discussed in Sect. 3, source-level analysis is often not sufficient, and binaries may contain leakages even though the original source code is constant-time. Therefore, we opted for a binary approach based on RISC–V-specific DBI for execution trace generation and Microwalk for leakage analysis.

5.1   Leakage Model

We adopt the leakage model as specified for Microwalk [51]: We supply the attacker with an implementation, a number of secret inputs and corresponding *execution traces*. An execution trace consists of a sequence of all executed instructions and accessed memory addresses, but does not contain actual processed data. The attacker also gets access to all public inputs and outputs. We consider the implementation constant-time if all traces are identical, i.e., when the attacker does not learn anything about the secret input by looking at a trace. In other words, in a constant-time program, the observed control flow and memory accesses are independent of the secret inputs.

This leakage model assumes a rather strong attacker, as the known side-channel attacks can only retrieve a fraction of the information expressed in a full execution trace. For example, cache attacks are limited to granularities of 32 or 64 bytes on most systems, and control flow tracking techniques like single-stepping only work in very specific scenarios. Due to the lack of suitable hardware, there has not yet been much work on side-channels for RISC–V. Thus, while we expect similar vulnerabilities on upcoming RISC–V processors as are already known for other architectures, sticking to a strong leakage model is the safest way forward. We only consider secret-dependent control flow and memory accesses that are architecturally reachable, so transient execution attacks are out-of-scope.

Implementation in Microwalk. Microwalk implements the above leakage model through a simple dynamic analysis pipeline, which generates secret inputs (called *test cases*), collects and preprocesses corresponding execution traces, and finally compares those traces with each other. If Microwalk finds a difference between two or more traces at a given code position, this difference is reported as *leakage*, as an attacker may exploit this difference to tell apart two or more secret inputs. If all traces are identical, the attacker does not learn anything about the underlying secret inputs, and the implementation is reported as non-leaking.

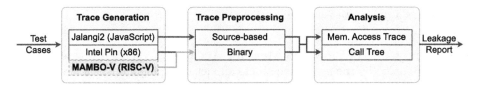

Fig. 3. Microwalk pipeline with a new trace generation module based on MAMBO–V. Each trace generation module may emit either source-based or binary execution traces, which are then preprocessed into a common trace format that can be parsed by all analysis modules.

5.2   Required Information

Microwalk uses a common generic execution trace format to run its analysis modules on, so we build a toolchain that collects RISC–V execution traces and converts them into Microwalk's format. Microwalk already offers two raw trace preprocessors, one for converting source-based execution traces from languages like JavaScript, and another one for binary traces from compiled code. While the binary trace preprocessor was originally written for x86, we found that its raw trace format is generic enough to also be used on other architectures. We thus only need to create a trace generator for RISC–V, that emits raw execution traces in the same format as the existing Intel Pin module (Fig. 3).

A raw binary execution trace from Microwalk's Intel Pin module combines the following information:

- taken/non-taken branches, with source and (if applicable) target address;
- memory accesses, with instruction address and accessed memory address;
- heap/stack allocation blocks, with start and end address;
- start and end addresses of the memory-mapped executable binaries.

We collect this data using a plugin for the MAMBO–V DBI framework.

5.3   MAMBO–V Trace Plugin

Interaction with the Target Program. In order to analyze a cryptographic primitive, the primitive has to be made available to the DBI framework. We follow Microwalk's approach by asking the user to supply a small function that receives a test case file with secret inputs and then calls the cryptographic primitive. Our MAMBO–V plugin registers a *function call* event callback for detecting execution of that function, so it can detect when test case execution starts and ends. This method has the advantage that we do not need to re-instrument the binary for each test case, but can reuse the existing instrumentation, which speeds up trace generation significantly. Before the first test case begins, we record a *trace prefix*, that contains initializations of all global objects that may be referenced during test case execution.

Recording Control Flow and Memory Accesses. When a test case begins, which is signaled by the respective event callback, our plugin opens a new binary trace file. We also register an instrumentation callback, which is called whenever a new basic block is instrumented. In this callback, we check each instruction for control flow and memory accesses, and add instrumentation to that instruction if necessary. The resulting instrumented code then writes to the trace file whenever the respective instruction is executed. To avoid tracing information outside our target functions, the plugin receives a list of binaries that should be traced.

Tracking Memory Allocations. Microwalk needs both a list of allocated heap memory blocks and the regions of the memory-mapped executables. To collect heap blocks, we register *function call* and *function return* event callbacks for the `malloc`, `calloc`, `realloc` and `free` functions, and log their parameters and return addresses. For the static memory regions, we hook into the *VM operation* event handler and extract the required information from `VM_MAP` events, which are triggered whenever a new ELF file is loaded.

6   Evaluation

To evaluate the performance of our toolchain and assess the current state of side-channel security on RISC–V, we analyze a number of frequently used cipher and signature functions for several popular libraries. We describe the experimental setup, analyze the performance of trace creation and analysis, and discuss and evaluate the discovered leakages. The results are summarized in Table 1.

6.1   Experimental Setup

As described in Sect. 3, we combine MAMBO–V with Microwalk to natively analyze the leakage of binaries on RISC–V. We record the traces with MAMBO–V on a Microchip PolarFire SoC FPGA Icicle Kit with four SiFive U54 cores featuring RV64GC. The trace analysis with Microwalk is executed on an AMD Ryzen 9 7950X with 16 cores.

Libraries. Due to its modular structure, the RISC–V architecture allows for a broad range of target applications, from small embedded devices to server CPUs. To reflect this, we chose to analyze WolfSSL [52] and Mbed TLS [31] as examples for libraries that support many architectures and that are optimized for the embedded market. OpenSSL [36] and GNU Nettle [16], on the other hand, are general purpose cryptography libraries that are used across different architectures and chip sizes. In addition, as an example of a library specifically written for RISC–V, we investigated SCL (SiFive Cryptographic Library) [44]. Finally, as a reference for constant-time implementations, we included libsodium [29].

Table 1. Result of leakage analysis of several cryptographic libraries on RISC–V. "Tr. CPU" shows the CPU time for generating the raw traces and "An. CPU" the CPU time for trace preprocessing and analysis. The columns "# Lkgs." and "# Uniq." show the total and unique number of detected leaking code lines.

Target	Type	Tr. CPU	An. CPU	# Lkgs	# Uniq.
WolfSSL [52] 5.5.4					
AES-ECB	cipher	1 sec	< 1 sec	157	157
AES-GCM	aead-cipher	2 sec	< 1 sec	493	184
ChaCha20-Poly1305	aead-cipher	< 1 sec	< 1 sec	0	0
Ed25519	signature	36 sec	< 1 sec	0	0
ECDSA (secp192r1)	signature	880 sec	7 sec	105	10
Mbed TLS [31] 3.3.0					
AES-ECB	cipher	2 sec	< 1 sec	68	68
AES-GCM	aead-cipher	4 sec	< 1 sec	216	76
ChaCha20-Poly1305	aead-cipher	7 sec	< 1 sec	0	0
OpenSSL [36] 3.0.0					
AES-ECB	cipher	115 sec	< 1 sec	52	52
AES-GCM	aead-cipher	117 sec	< 1 sec	166	60
ChaCha20-Poly1305	aead-cipher	117 sec	< 1 sec	0	0
Ed25519	signature	556 sec	4 sec	0	0
ECDSA (secp192r1)	signature	3128 sec	30 sec	1647	284
GNU Nettle [16] 3.8.1 with GMP [15] 6.2.1					
AES-ECB	cipher	2 sec	< 1 sec	32	32
AES-GCM	aead-cipher	3 sec	< 1 sec	108	40
ChaCha20-Poly1305	aead-cipher	2 sec	< 1 sec	0	0
Ed25519	signature	104 sec	4 sec	0	0
SCL - SiFive Cryptographic Library [44] 20.08.00					
ECDSA (secp256r1)	signature	102 sec	< 1 sec	5	2
libsodium [29] 1.0.18					
ChaCha20-Poly1305	aead-cipher	2 sec	< 1 sec	0	0
Ed25519	signature	12 sec	< 1 sec	0	0

Analyzed Primitives. We wrote analysis wrappers for AES-ECB, the authenticated encryption schemes AES-GCM and ChaCha20-Poly1305, and the signature algorithms Ed25519 and ECDSA (curve `secp192r1`; `secp256r1` for SCL). The wrappers initialize the necessary environment and call the target functions, if supported by the respective library. We skipped the ECDSA implementations in GNU Nettle and Mbed TLS, as those are comparably slow and thus lead to traces which exceed the limited resources of our evaluation platform.

All libraries and target wrappers were cross-compiled with the RISC–V GNU Compiler Toolchain 12.2.0 [42] for RV64GC and ISA specification 2.2. We built all libraries with default options and appropriate additional security flags as stated in their documentation. All libraries except OpenSSL are built with optimization level `-O2`. OpenSSL was built with optimization level `-O3`.

Test Cases. We generated 16 test cases for each primitive by creating 16 random keys, and supplied these test cases to the target function. Since Microwalk measures differences in the execution traces, any other input outside the test cases must be kept constant to avoid false positives. Therefore, inputs such as initialization vectors were set to fixed values. Random values like the ephemeral key in ECDSA were generated by custom test case-dependent RNGs. We opted for using smaller key sizes, as the cryptographic procedures are invariant of the key size, and larger key sizes increase the resource consumption of the leakage analysis without uncovering further vulnerabilities [51].

6.2 Performance Results

The performance of the side-channel analysis on RISC–V depends on the time required for tracing the target function and analyzing the traces. The runtime for all targets is summarized in Table 1.

Tracing. The duration of tracing 16 executions for each target is inherently constrained by the limited performance of the SiFive U54 core. For the symmetric ciphers and Ed25519, the tracing took at most a few minutes, which suggests that our toolchain is suitable for everyday use on a developer's computer. With newer and more performant RISC–V cores, the tracing time should further decrease.

One outlier is OpenSSL, where a majority of the tracing time was spent in the library initialization, which is mostly irrelevant for the leakage analysis. To reduce this overhead, the developer could disable most features when compiling the library for vulnerability evaluation and target low-level functions.

Analysis. With one exception, the trace preprocessing and analysis of nearly all targets took less than 5 s. The fast analysis allows for frequent execution of any test. The outlier, ECDSA for OpenSSL, was slowed down by preprocessing the huge traces, so optimizing the tracing time should fix this as well.

6.3   Vulnerabilities

The leakage analysis for the chosen popular libraries shows many vulnerabilities across the board, except for libsodium which only implements a limited number of ciphers and signature algorithms that allow for an implementation with better resistance against timing attacks by design. Indeed, all analyzed implementations of ChaCha20-Poly1305 and Ed25519 are constant-time. We summarize the results in Table 1 in the columns "# Lkgs." (total leakages) and "# Uniq." (unique leakages). An instruction or function can be called or reached from multiple contexts, thus potentially leaking different secrets with varying leakage severity. Therefore, we also count unique occurrences of leaking instructions.

In-depth analysis of the libraries showed that most provide specific assembly implementations for x86 and other architectures that use constant-time primitives. For RISC–V though, due to lack of specifically optimized implementations, the libraries fell back to default ones, which often turned out to be non-constant-time, even when using the hardening flags specified in the documentation.

Symmetric Ciphers. All analyzed AES-ECB implementations leak secret information through their timing behavior. The examined libraries do not provide RISC–V-specific code, but fall back to their default C/C++ implementations, which use either T-table or S-box lookups for AES encryption and round key generation. Previous work has shown that table lookups are exploitable by timing measurements [7]. The number of unique leakages varies between the different libraries depending on whether the encryption rounds are unrolled and how the final step is scheduled. After informing the OpenSSL developers that we found several leakages in the default AES-ECB implementation, we were pointed to an undocumented compiler flag that enables an alternative AES implementation, which we verified to be constant-time. However, they also stated that the flag leads to a 95% performance loss, which is why it is not enabled by default.

The authenticated encryption algorithm AES-GCM builds upon the same primitives as AES-ECB and thus also shows the same table lookup leakage for the encryption step. In addition, the GCM mode adds authentication through computation of a GHASH, which involves encryption of a 128-bit string of zeros and the IV. The result of the latter encryption is used for the final computation of the authentication data. The multiplication used for the GHASH is implemented with a hash lookup table, where the accessed index depends on the current ciphertext and the hash value of the previous block.

We compared the leakage result of AES-GCM on RISC–V for the libraries OpenSSL and Mbed TLS against the analysis on x86. While the RISC–V binaries contain many leakages as explained above, we observed no leakages for x86 binaries. The x86 implementations use the AES-NI hardware extension for encryption and the `clmul` extension for computation of the GHASH. Until such extensions are available for RISC–V, cryptographic libraries must feature constant-time software implementations. For WolfSSL, we learned during disclosure that there is a `GCM_SMALL` flag, which enables a non-table-based GHASH implementation.

While designed (and documented) primarily for small code size, we found that it is constant-time and thus a secure alternative for the default implementation.

Asymmetric Signature Algorithms. None of the analyzed implementations of Ed25519 shows any non-constant-time behavior, emphasizing its inherent resistance against timing attacks, even though there are no specific assembly implementations for RISC–V. However, we found leakage for all analyzed implementations of ECDSA, especially in the implementation from OpenSSL. Even the specially crafted RISC–V implementation from SCL reveals non-constant-time behavior, though the library is not yet deemed production-ready. Despite the high number of potential vulnerabilities, we found that all analyzed ECDSA implementations use blinding, rendering the discovered leakages likely unexploitable.

7   Discussion and Future Work

Limitations of Microwalk. As we base our analysis on Microwalk, we inherit some of its limitations. Currently, Microwalk only supports deterministic implementations. Thus, all entropy must come from the secret inputs. While this scenario works well with symmetric and constant-time asymmetric cryptographic primitives, it has some issues with blinded implementations which obscure the computation by randomizing the input parameters. Disabling the randomness is not sufficient either, as this would just expose leakages which are normally obscured by blinding. As a solution, Microwalk should be extended to support randomized implementations. Another limitation of Microwalk's analysis algorithm is the possibility of several small leakages higher up in the call chain hiding leakages further down, though we did not observe this during our evaluation. Finally, Microwalk's dynamic approach heavily depends on the coverage. While it was found that few random test cases usually suffice [48,51], the user should check that all relevant code locations have been reached.

Other Applications of MAMBO–V. While we used MAMBO–V for generating execution traces, the tool is far more versatile. The plugin API supports a variety of different callbacks, making it on par with other widely-used frameworks like Intel Pin. For example, new plugins can aid with control-flow checks or help in bug detection. The broad similarities to ARM allow reusing analysis code originally written for MAMBO with little adjustments.

Leakage Analysis on ARM. The proximity of RISC–V and ARM suggests that the MAMBO–V trace generator plugin can be ported to the original MAMBO implementation with little adjustments. With that plugin, one could generate execution traces from ARM binaries, and analyze these traces for side-channel vulnerabilities using Microwalk, yielding a dynamic leakage analysis toolchain for ARM. Thus, our toolchain comprising a tracer plugin and Microwalk provides a solid basis for fast and accurate side-channel leakage analysis on various systems.

8   Related Work

Analysis of Code on Intermediate Representations. Instead of instrumenting code natively, the machine code can be lifted to a generic intermediate representation. This approach is taken by the ongoing RISC–V port [38] of the heavyweight instrumentation framework Valgrind [35] and the full-system emulator QEMU [6], which do an emulated analysis of RISC–V instructions on the intermediate representations of the respective framework. Thereby, it is possible to re-use existing analysis tools like memory leaks detection or call graphs. Apart from that, the whole system reverse engineering tool PANDA [13] provides a way to capture an execution trace, replay it afterwards and combine it with extensive analysis through different plugins. However, emulated analysis meets a different objective than analyzing architecture-specific leakage, as the leakage may be hidden during lifting to the intermediate representation. Furthermore, the emulators impose a very high overhead and are too resource-consuming to use them in restricted environments or for an efficient analysis with Microwalk.

Side-Channel Analysis. Side-channel attacks on RISC–V are receiving growing attention by security research. Apart from the timing side-channels we analyze in this work, there have been efforts to secure RISC–V implementations against leakage through power side-channels [32]. Further, electromagnetic leakage builds the basis for a successful fault attack in [34], showing that manifold leakage channels need to be addressed. As some RISC–V systems also support out-of-order execution, they are susceptible to Spectre [25] attacks [17,27]. Recently, it was shown that data can be leaked from speculative execution through cache attacks [28]. The vulnerability to Spectre-style attacks further motivates the development of a framework to automatically detect timing side-channels in software, because apart from direct exploitation, the timing differences can also be used as a way to leak speculatively accessed secrets.

Hardware-Based Countermeasures. A RISC–V working group developed a number of extensions intended for secure cryptography, which were ratified in 2022 [41]. This includes hardware-acceleration for symmetric encryption and hash functions, but also the *Zkt* extension, which specifies constant-time properties for certain instructions. If a vendor implements the Zkt extension, certain arithmetic instructions are guaranteed to have data-independent execution time. However, solely instruction-based approaches are insufficient, as most vulnerabilities are caused by higher-level data-dependent behavior. Yu et al. propose support for oblivious memory accesses, which would block most timing side-channels [54] and thus go far beyond simply avoiding data-dependent instruction latency like in the Zkt extension. With hardware-integrated fully automated Boolean masking [45], hardly any software-level precautions need to be taken against power side-channels. To protect against data leakages in ALU, memory and memory interfaces, INVITED [32] uses state-of-the-art masking techniques.

However, these hardware mechanisms are always applied, not only for secret inputs, making the solutions potentially inefficient for workloads where only a small fraction of all executed instructions is truly security-critical. Moreover, in a cloud scenario, the clients have limited control about the hardware actually used, making secure software implementations indispensable.

9   Conclusion

In this paper, we have presented the first comprehensive side-channel analysis for implementations of cryptographic primitives on RISC–V. We have shown that some of the most popular open-source cryptographic libraries lack proper side-channel resistance on RISC–V. For our work, we have studied the requirements for leakage detection on RISC–V and designed a thorough approach to incorporate all requirements into a mature side-channel analysis framework that we have extended with all necessary building blocks. We have based our analysis toolchain on Microwalk and augmented the framework with the necessary RISC–V specific tracing capabilities by implementing the DBI tool MAMBO–V. Our evaluation pinpoints several potentially exploitable leakages that should be fixed by the developers and emphasizes the need for complete and precise side-channel analysis capabilities on RISC–V to pave the way for secure computations on shared RISC–V hardware in the cloud.

Acknowledgements. We thank the library maintainers for the smooth disclosure process, and the reviewers and our shepherd for their helpful comments and suggestions. This work has been supported by DFG under grants 427774779 and 439797619, and by BMBF through projects ENCOPIA and PeT-HMR.

References

1. Aciiçmez, O., Brumley, B.B., Grabher, P.: New results on instruction cache attacks. In: Cryptographic Hardware and Embedded Systems, CHES 2010, 12th International Workshop (2010)
2. Aciiçmez, O., Koç, Ç.K., Seifert, J.: Predicting secret keys via branch prediction. In: Topics in Cryptology - CT-RSA (2007)
3. Almeida, J.B., Barbosa, M., Barthe, G., Dupressoir, F., Emmi, M.: Verifying constant-time implementations. In: 25th USENIX Security Symposium (2016)
4. Asanović, K., Patterson, D.A.: Instruction sets should be free: the case for RISC-V. EECS Dpt., Univ. of CF, Berkeley, Technical report, UCB/EECS-2014-146 (2014)
5. Bao, Q., Wang, Z., Li, X., Larus, J.R., Wu, D.: Abacus: precise side-channel analysis. In: 43rd IEEE/ACM International Conference on Software Engineering, ICSE (2021)
6. Bellard, F.: QEMU, a fast and portable dynamic translator. In: Proceedings of the FREENIX Track: 2005 USENIX Annual Technical Conference (2005)
7. Bernstein, D.J.: Cache-Timing Attacks on AES (2005)
8. Brotzman, R., Liu, S., Zhang, D., Tan, G., Kandemir, M.T.: CaSym: cache aware symbolic execution for side channel detection and mitigation. In: IEEE Symposium on Security and Privacy, S&P (2019)

9. Bruening, D., Garnett, T., Amarasinghe, S.P.: An infrastructure for adaptive dynamic optimization. In: 1st IEEE/ACM International Symposium on Code Generation and Optimization (CGO) (2003)
10. Daniel, L., Bardin, S., Rezk, T.: BINSEC/REL: symbolic binary analyzer for security with applications to constant-time and secret-erasure. ACM Trans. Privacy Secur. (2022)
11. Dessouky, G., Frassetto, T., Sadeghi, A.: HybCache: hybrid side-channel-resilient caches for trusted execution environments. In: 29th USENIX Security Symposium (2020)
12. Dinesh, S., Burow, N., Xu, D., Payer, M.: RetroWrite: statically instrumenting COTS binaries for fuzzing and sanitization. In: IEEE Symposium on Security and Privacy, S&P (2020)
13. Dolan-Gavitt, B., Hodosh, J., Hulin, P., Leek, T., Whelan, R.: Repeatable reverse engineering with PANDA. In: Proceedings of the 5th Program Protection and Reverse Engineering Workshop, PPREW@ACSAC (2015)
14. Doychev, G., Feld, D., Köpf, B., Mauborgne, L., Reineke, J.: CacheAudit: a tool for the static analysis of cache side channels. In: Proceedings of the 22nd USENIX Security Symposium (2013)
15. GMP: https://ftp.gnu.org/gnu/gmp/. Accessed 24 Jan 2023
16. GNU Nettle: https://git.lysator.liu.se/nettle/nettle. Accessed 24 Jan 2023
17. Gonzalez, A., Korpan, B., Zhao, J., Younis, E., Asanovic, K.: Replicating and mitigating spectre attacks on an open source RISC-V microarchitecture. In: Third Workshop on Computer Architecture Research with RISC-V (CARRV) (2019)
18. Gorgovan, C., D'Antras, A., Luján, M.: MAMBO: a low-overhead dynamic binary modification tool for ARM. ACM Trans. Archit, Code Optim (2016)
19. Gras, B., Razavi, K., Bos, H., Giuffrida, C.: Translation leak-aside buffer: defeating cache side-channel protections with TLB attacks. In: 27th USENIX Security Symposium (2018)
20. He, S., Emmi, M., Ciocarlie, G.F.: ct-fuzz: fuzzing for timing leaks. In: 13th IEEE International Conference on Software Testing, Validation and Verification, ICST (2020)
21. Inci, M.S., Gülmezoglu, B., Irazoqui, G., Eisenbarth, T., Sunar, B.: Cache attacks enable bulk key recovery on the cloud. In: Cryptographic Hardware and Embedded Systems - CHES (2016)
22. Irazoqui, G., Eisenbarth, T., Sunar, B.: S$A: a shared cache attack that works across cores and defies VM sandboxing - and its application to AES. In: IEEE Symposium on Security and Privacy, S&P (2015)
23. Jancar, J., et al.: "They're not that hard to mitigate": what cryptographic library developers think about timing attacks. In: 43rd IEEE Symposium on Security and Privacy, S&P (2022)
24. Kaufmann, T., Pelletier, H., Vaudenay, S., Villegas, K.: When constant-time source yields variable-time binary: exploiting curve25519-donna built with MSVC 2015. In: Cryptology and Network Security - 15th International Conference, CANS (2016)
25. Kocher, P., et al.: Spectre attacks: exploiting speculative execution. In: IEEE Symposium on Security and Privacy, S&P (2019)
26. Kumar, V.B.Y., et al.: Towards Designing a Secure RISC-V System-on-Chip: ITUS. J. Hardw. Syst. Secur. (2020)

27. Le, A.T., Dao, B.A., Suzaki, K., Pham, C.K.: Experiment on replication of side channel attack via cache of RISC-V berkeley out-of-order machine (BOOM) implemented on FPGA. In: Fourth Workshop on Computer Architecture Research with RISC-V (CARRV) (2020)
28. Le, A.T., Hoang, T.T., Dao, B.A., Tsukamoto, A., Suzaki, K., Pham, C.K.: A cross-process spectre attack via cache on RISC-V processor with trusted execution environment. Comput. Electr. Eng. (2023)
29. libsodium: https://github.com/jedisct1/libsodium. Accessed 24 Jan 2023
30. Luk, C., et al.: Pin: building customized program analysis tools with dynamic instrumentation. In: Proceedings of the ACM SIGPLAN Conference on Programming Language Design and Implementation (2005)
31. Mbed-TLS: https://github.com/Mbed-TLS/mbedtls. Accessed 24 Jan 2023
32. Mulder, E.D., Gummalla, S., Hutter, M.: Protecting RISC-V against side-channel attacks. In: Proceedings of the 56th Annual Design Automation Conference, DAC (2019)
33. Nasahl, P., Schilling, R., Werner, M., Mangard, S.: HECTOR-V: a heterogeneous CPU architecture for a secure RISC-V execution environment. In: ASIA CCS: ACM Asia Conference on Computer and Communications Security (2021)
34. Nashimoto, S., Suzuki, D., Ueno, R., Homma, N.: Bypassing isolated execution on RISC-V using side-channel-assisted fault-injection and its countermeasure. IACR Trans. Cryptogr. Hardw. Embed. Syst. (2022)
35. Nethercote, N., Seward, J.: Valgrind: a program supervision framework. In: Third Workshop on Runtime Verification, RV@CAV (2003)
36. OpenSSL: https://github.com/openssl/openssl. Accessed 24 Jan 2023
37. Osvik, D.A., Shamir, A., Tromer, E.: Cache attacks and countermeasures: the case of AES. In: Topics in Cryptology - CT-RSA (2006)
38. Pavlu, P.: Valgrind RISC-V Port. Free and Open Source Software Developers' European Meeting (FOSDEM) (2022). https://github.com/petrpavlu/valgrind-riscv64
39. QBDI: QuarkslaB Dynamic binary Instrumentation. https://qbdi.quarkslab.com/
40. Reparaz, O., Balasch, J., Verbauwhede, I.: Dude, is my code constant time? In: Design, Automation & Test in Europe Conference & Exhibition, DATE (2017)
41. RISC-V: RISC-V Cryptography Extensions Volume I. https://github.com/riscv/riscv-crypto/releases/tag/v1.0.1-scalar. Accessed 24 Jan 2023
42. RISC-V Software Collaboration: RISC-V GNU Compiler Toolchain. https://github.com/riscv-collab/riscv-gnu-toolchain. Accessed 24 Jan 2023
43. Saileshwar, G., Qureshi, M.K.: MIRAGE: mitigating conflict-based cache attacks with a practical fully-associative design. In: 30th USENIX Security Symposium (2021)
44. SiFive Cryptographic Library (SCL): https://github.com/sifive/scl-metal. Accessed 24 Jan 2023
45. Stangherlin, K., Sachdev, M.: Design and implementation of a secure RISC-V microprocessor. IEEE Trans. Very Large Scale Integr. Syst. (2022)
46. Wang, S., Bao, Y., Liu, X., Wang, P., Zhang, D., Wu, D.: Identifying cache-based side channels through secret-augmented abstract interpretation. In: 28th USENIX Security Symposium (2019)
47. Waterman, A.: Design of the RISC-V Instruction Set Architecture. Ph.D. thesis, University of California, Berkeley, USA (2016)
48. Weiser, S., Zankl, A., Spreitzer, R., Miller, K., Mangard, S., Sigl, G.: DATA-differential address trace analysis: finding address-based side-channels in binaries. In: 27th USENIX Security Symposium (2018)

49. Werner, M., Unterluggauer, T., Giner, L., Schwarz, M., Gruss, D., Mangard, S.: ScatterCache: thwarting cache attacks via cache set randomization. In: 28th USENIX Security Symposium (2019)
50. Wichelmann, J., Moghimi, A., Eisenbarth, T., Sunar, B.: MicroWalk: a framework for finding side-channels in binaries. In: Proceedings of the 34th Annual Computer Security Applications Conference, ACSAC (2018)
51. Wichelmann, J., Sieck, F., Pätschke, A., Eisenbarth, T.: Microwalk-CI: practical side-channel analysis for JavaScript applications. In: Proceedings of the ACM SIGSAC Conference on Computer and Communications Security, CCS (2022)
52. WolfSSL: https://github.com/wolfSSL/wolfssl. Accessed 24 Jan 2023
53. Yarom, Y., Falkner, K.: FLUSH+RELOAD: a high resolution, low noise, l3 cache side-channel attack. In: Proceedings of the 23rd USENIX Security Symposium (2014)
54. Yu, J., Hsiung, L., Hajj, M.E., Fletcher, C.W.: Data oblivious ISA extensions for side channel-resistant and high performance computing. In: 26th Annual Network and Distributed System Security Symposium (NDSS) (2019)

The Finger in the Power: How to Fingerprint PCs by Monitoring Their Power Consumption

Marina Botvinnik[1,4]([✉]), Tomer Laor[1], Thomas Rokicki[2], Clémentine Maurice[3], and Yossi Oren[1,4]

[1] Ben-Gurion University of the Negev, Beersheba, Israel
botvinnik95@gmail.com
[2] Univ Rennes, CNRS, IRISA, Rennes, France
[3] Univ Lille, CNRS, Inria, Lille, France
[4] Intel Corporation, Santa Clara, USA

Abstract. Power analysis has long been used to tell apart different instructions running on the same machine. In this work, we show that it is also possible to use power consumption to tell apart *different machines running the same instructions*, even if these machines have entirely identical hardware and software configurations, and even if the power consumption measurements are carried out using low-rate software-based methods. We collected an extended dataset of power consumption traces from 291 desktop and server systems, spanning multiple processor generations and vendors (Intel and AMD). After analyzing them, we discovered that profiling the power consumption of individual assembly instructions makes it possible to create a fingerprinting agent that can identify individual machines with high accuracy. Our classifier approaches its peak accuracy after less than 10 instructions, meaning that the fingerprint can take a very short time to capture. We analyzed the stability of the fingerprint over time and discovered that, while it remains relatively stable, it is significantly affected by temperature changes. We also carried out a proof-of-concept evaluation using portable WebAssembly code, showing that our method can still be applied, albeit at a reduced accuracy, without using native instructions for the profiling step. Our method depends on the ability to measure power, which is currently restricted to high-privileged "ring 0" code on modern PCs. This limits the current use of our method to defense-only settings, such as strengthening authentication or anti-counterfeiting. Our tools and datasets are publicly released as an open-source repository. Our work highlights the importance of protecting power consumption measurements from unauthorized access.

Keywords: Side Channel · Fingerprinting · PUF · WebAssembly

1 Introduction

As surprising as it may seem, individual copies of mass-manufactured computing devices are never completely identical. Minuscule variations introduced during

---

M. Botvinnik, T. Laor and T. Rokicki contributed equally to this paper.

© The Author(s), under exclusive license to Springer Nature Switzerland AG 2023
D. Gruss et al. (Eds.): DIMVA 2023, LNCS 13959, pp. 24–45, 2023.
https://doi.org/10.1007/978-3-031-35504-2_2

the hardware manufacturing process result in differences in the behaviors of elements that form an integrated circuit (IC), including storage, logic, and communication.

This interesting phenomenon is investigated by researchers in the field of fingerprinting. Fingerprinting extracts the unique attributes of each device and uses them to differentiate each device from other similar devices. Thus, a unique physical fingerprint should be able to differentiate one device from the others even when the software stack is identical on all devices (e.g., the same operating system and software are installed with the same versions), and all hardware components are the same model (e.g., same CPU and DRAM models). Several hardware features or components have been used for physical fingerprinting, such as DRAM [34,38], SRAM [10], and GPUs [19]. While it was originally be used in the context of ICs, physical fingerprinting is now used in a variety of contexts that require some identification, such as IoT devices [26], FPGA boards in the cloud [39], and mobile devices using sensors [7,11]. On the defensive side, fingerprinting can play a major role in multi-factor authentication [7,20], access control [3], and even anti-counterfeiting [5]. On the attacking side, it can be used for tracking devices and users without their consent [21].

In this work, we turn our attention to another potential source of fingerprinting information – the *power consumption* of a PC as it executes different instructions. The power consumption of CMOS devices, such as computers, varies depending on the instructions executed or the data processed [25]. This effect is actively being used by the security research community to carry out *power analysis attacks* – attacks which discover secrets about the internal state of various computing devices by analyzing their power consumption – ever since the publication of the seminal work of Kocher et al. in 1999 [16]. While traditional power analysis attacks require physical access to the device under test (DUT), a growing body of works has explored methods of running software-only power analysis attacks, relying on alternative methods for measuring power consumption launched remotely [4,23,35].

In parallel to the work done by the security research community, the performance engineering research community also has an interest in power consumption measurements, since the limited power and thermal budgets of computer systems is one of the main factors determining how fast code can be run. In an interesting work coming from this community, von Kistowski et al. [15] noted that seemingly-identical machines have different power consumption when performing identical tasks. They found that the power consumption for common benchmarks run on commercially identical processors can vary between computers by as much as 29.6% for an idle CPU and 19.5% at full load. While von Kistowski et al. considered their observation as a negative result, highlighting the challenge of uncertainty when dealing with benchmarks, we were motivated to investigate whether this variation can actually serve as a fingerprinting mechanism that can identify individual PCs.

In this work, we show that this difference in power consumption among identical computers can indeed be used to distinguish among them with high accuracy. In particular, we show how we can distinguish between identical machines at an accuracy of up to 65 times higher than random guessing. While it is currently limited by restrictions on user-mode power consumption measurements related to the PLATYPUS disclosures [23], the fingerprint is quite stable in time and takes a reasonable time to capture.

We evaluated our method on several sets of identical computers. We also show that this method can be reproduced by using web client-side workload, in particular by using WebAssembly instructions. Although the fingerprinting still requires a native access to read the power consumption, the web fingerprinting allows to improve the experiment's portability as well as greatly reducing the code base.

Contributions. The main contributions are as follows:

- We show that it is possible to create a fingerprint based on power consumption of the CPU. We evaluate our methods on 291 desktop and server systems, spanning multiple processor generations and vendors (Intel and AMD), and show that it consistently delivers accuracy significantly higher than the base rate (76% for a set of 17 Core i5-4590 desktops, 59% for a set of 71 Xeon E5-2630 servers, 55% for a set of 123 Xeon Gold 5220, 89% for a set of Xeon Gold 6130, and even 91% on 7 AMD EPYC 7301).
- We evaluate the influence of CPU temperature and time drift over power-consumption based fingerprint. We demonstrate that while time drift decreases the accuracy of the fingerprint, taking into account the CPU temperature increases its accuracy.
- We show a proof of concept of web-based fingerprint based on power consumption, yielding 35% accuracy on a set of 17 computers, showing that power-consumption fingerprinting can also be applied from a high-level portable languages, and be oblivious to the microarchitecture.

Our work presents a fingerprinting vector that can increase the accuracy of existing defensive fingerprinting systems. It also serves as another warning against providing unrestricted access to computer power consumption measurements.

2   Background

CPU Fingerprinting. A fingerprint is often composed of one or several attributes creating a unique identifier. The quality of such an attribute is evaluated with two significant properties. The first property is *uniqueness*: A fingerprint's end goal is uniquely identifying a user or device. To that extent, a perfect attribute would be unique. However, such attributes are hard to encounter. The second is *stability*: Changes in an attribute can break the fingerprint and prevent users' identification. A stable attribute does not vary significantly with time or can be linked to previous iterations. In that regard, hardware attributes are interesting as they offer high stability, as users rarely change hardware components. They are thus valuable in strengthening more volatile software-based fingerprints.

Hardware attributes can fall into either of two categories. *Discrete attributes* are classified in pre-determined categories, such as the number of physical cores [40] or CPU generation [29]. As many users share the same hardware model, these attributes do not yield a high uniqueness, but their identification is often stable. On the contrary, *continuous attributes* exploit side effects of manufacture to create an attribute unique to an iteration of the hardware component. These attributes are often complex to measure as they do not fall into pre-determined categories and yield a high uniqueness.

Power Analysis. The power consumption of CPUs is data-dependent, *i.e.*, it varies based on the instructions executed or the data processed. Power analysis is a type of side channel extracting information from these slight differences. Kocher et al. [16] introduced differential power analysis: by physically measuring how the power consumption varies at a fixed point in a function's execution, an attacker can infer the data processed. They use differential power analysis to extract DES private keys. This side channel has been expanded and modeled by Messerges et al. [27]. Mangard et al. [25] proposed an overview of power-consumption attacks and techniques to improve the signal. All these hardware-based power side channels require physical access to the device and specialized hardware, e.g., an oscilloscope. More recently, these power side channels have been explored in a pure software implementation, without physical access to the device [22,23]. These attacks leverage software interfaces,e.g., Intel's RAPL, allowing a user to get power consumption and CPU temperature feedback at a high frequency.

WebAssembly. WebAssembly is a bytecode-like language of the web, designed for client-side computations *i.e.*, executed directly in the users' browsers, in sandboxed environments. WebAssembly can be compiled directly from other languages, e.g., C or Rust, or written in the `wat` text format, an assembly-like representation of the binary code. WebAssembly standards are currently composed of up to 256 instructions, offering more fine-grained control than JavaScript. It is built in a typed stack-machine model.

3   Fingerprinting Model

In the model we use for this work, we assume a fingerprinting agent capable of running short code sequences on the device under test (DUT) and measuring their power consumption. The agent's goal is to distinguish between n computers, labeled $c_1 \cdots c_n$, using power consumption data as the classification feature, as presented in Fig. 1. The system should work even if all n computers have identical hardware and software stacks.

The fingerprinting process begins by selecting a group of m assembly-language instructions, labeled $i_1 \cdots i_m$. We evaluate two settings for this model. In the first, described in Sect. 4.2, we assume the assembly-language instructions are written in native code. In the second, described in Sect. 4.3, we assume the assembly-language instructions are delivered in portable form as WebAssembly

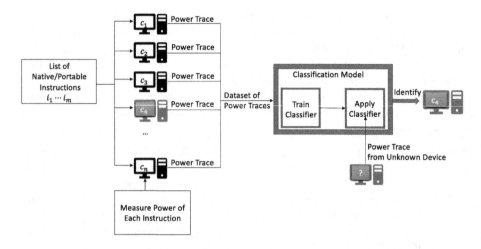

Fig. 1. Fingerprinting devices using power consumption.

instructions, and then compiled on the fly into native code by the DUT's web browser. In the next step, the agent measures the power consumption of each individual instruction using a software-based method, as described below. The agent also collects some additional data, including the time taken to execute the instruction and the core temperature at the time of measurement. This process is repeated for each instruction in the set to be measured, ultimately obtaining a *trace* of power consumption measurements of length m.

Once a trace is defined, our problem follows a standard classification workflow: In an offline *profiling* step, the agent captures multiple power consumption traces from multiple computers. Next, the power traces are used to construct a machine-learning classifier. Then, in an online *fingerprinting* step, the agent captures a single power trace from an unknown computer, and must use the classifier constructed in the offline phase to correctly identify which of the computers emitted this unlabeled trace.

Key Performance Indicators. We can evaluate our fingerprinting system's quality using multiple parameters. First and foremost are the fingerprint's *uniqueness and stability*, corresponding to its ability to identify individual machines accurately and consistently over time. Additional parameters are the *speed* of the fingerprint collection process and its *compatibility* with multiple types of hardware from multiple vendors and architectural generations.

4   Methodology

State of the art of hardware fingerprinting mechanisms focuses on detecting static CPU properties, such as the cache size or micro-architectural generation [29, 40] or on the relative speed of the machine's underlying components [19, 32]. This work, in contrast, focuses on the power consumption of the CPU – we assume that, due to slight manufacturing differences in the hardware, the power consumption is slightly different between each device. We would like to empirically demonstrate that this information is enough to significantly improve the fingerprint accuracy beyond the base rate of a naive classifier choosing one of the devices at random.

4.1 Fingerprinting Process Overview

As presented in Sect. 3, the goal of the classifier is to distinguish between n computers, labeled $c_1 \cdots c_n$, using the power consumption of each computer as the classification feature. We repeat the trace collection process ℓ times for each computer. Our dataset thus contains a total of $(\ell \times m \times n)$ power measurements. After gathering the dataset, we build a classification model, as described in Sect. 5.1. The model receives as input a single trace from one of the machines in the dataset, and predicts which machine created this power trace. The power trace will be collected in the same process as the entire dataset, hence, it will be a list of power consumption measurements of size m. To limit the noise in our measurements, we execute all instructions on the same physical core, ensuring no other processes are running on this core.

Measuring Power Consumption. While the instructions to be profiled are all unprivileged, ring 3 instructions, our model also assumes that the fingerprinting agent is capable of measuring the average power consumption of the device under test, as well as its temperature. Software applications running on Intel and AMD processors can monitor power consumption, without requiring external hardware, by accessing a model-specific register (MSR) named Running Average Power Limit (RAPL). RAPL is a hardware feature designed to monitor and control the system's overall power consumption. It includes an interface for reporting the accumulated energy consumption of various power domains, including the CPU, its attached DRAM, and other components such as the on-chip GPU [14]. A similar MSR also exists for AMD processors, with similar capabilities. Linux offers an easy-to-use interface to the RAPL registers through the /sys/class/powercap/intel-rapl/intel-rapl:0/intel-rapl:0:0/energy_uj virtual file system using PP0 domain, allowing them to be read using high-level scripting languages. RAPL-based measurement is performed in practice by sampling the system's accumulated energy consumption, executing the workload, and finally sampling the accumulated energy consumption once again, and then storing the difference between the final and initial energy measurements.

The main limitation of using RAPL is its required privilege level. Starting in October 2020, following the revelations of Lipp et al. [22,23], access to the RAPL interface was restricted to privileged processes. Consequently, the agent we describe must be trusted by the system owners being fingerprinted, challenging our ability to use the agent in an offensive setting. We further note that when the CPU is running in "Filtered RAPL" mode [12], RAPL readings are passed through a filter which reduces their update frequency and adds some

Table 1. Evaluated system specifications.

Name	Vendor	CPU Type	Node Count	μ-Arch.	Year
DESK-4590	Intel	Core i5-4590	17	Haswell	2014
SRV-2630-V3	Intel	Xeon E5-2630 v3	71	Haswell	2014
SRV-2630L-V4	Intel	Xeon E5-2630L v4	46	Broadwell	2016
SRV-6130	Intel	Xeon Gold 6130	27	Skylake	2017
SRV-5220	Intel	Xeon Gold 5220	123	Cascade Lake	2019
SRV-AMD	AMD	EPYC 7301	7	Zen 1	2017

random noise. This mode, which may affect our method's effectiveness, is currently engaged only when SGX is enabled, but may be extended to other settings in the future. We propose some workarounds to this limitation in Sect. 6.2.

4.2 Native Code Setup

The systems we evaluated are listed in Table 1. We chose six evaluation sets that vary in their characteristics. The DESK-4590 set consists of desktop machines, whereas the rest of the sets (SRV-) are servers located in the Grid'5000 testbed. The systems represent micro-architectural designs spanning multiple processor generations and multiple vendors. We note that the CPUs we evaluated do not support Intel's Software Guard Extension (SGX) feature which, as noted above, may limit the effectiveness of RAPL readings when it is enabled.

Grid'5000 Environment. One of the challenges in evaluating fingerprinting schemes for desktop computers is the difficulty of obtaining multiple systems with identical software and hardware configurations. Obviously, any external difference in the hardware, or in their environments, may be reflected onto the traces and may skew the measured performance of the fingerprinting algorithm. Previous works have attempted to address this challenge by using university computer classrooms, or by crowd-sourcing the experiment and clustering the data into multiple groups after it is collected based on other features [19,32]. In this work we present a novel approach that further reduces the risk of external factors affecting the fingerprint. We collaborated with Grid'5000, a large-scale parallel and distributed computing testbed. Grid'5000 has several clusters consisting of multiple hardware nodes. Each cluster node is identically configured and located in the same data center, ensuring that environmental variation is tightly controlled. Furthermore, since these systems are typically used for distributed computing tasks, there is less chance that software installed on one particular node affects measurements.

Our code template is based on Gras et al. [8]. The entire x86-64 set, including its optional instruction set extensions, consists of more than 16,000 different instructions and instruction variations. To make the experiment practical, we selected a representative sample of 455 instructions. We chose one of the instruction sets of Gras et al. [8] for our evaluation, in particular instructions that execute on CPU ports 0, 1, and 5 that were used by Gras et al. in their research for port contention. Each trace contains the power consumption of 455 instructions, with each instruction considered a feature. Although other instructions can be considered (as we mention in Sect. 6.2), we prioritized reproducibility over performance when selecting the instruction set and writing the data collection code. The measurement process is pinned and executed on one core, while the other pipeline code is pinned to another core to avoid interferences.

4.3 Portable Code Proof of Concept Setup

Web client-side computations often allow more portability as they reduce the code base and are adaptable, by design, to most systems that can run browsers. The user downloads the script from the server and runs it automatically. Web-based fingerprinting would render the process more portable, significantly reducing the code base of the experiments and making it highly adaptive to different operating systems or browsers. We propose a proof of concept of web-based power fingerprinting. This fingerprinting is built around WebAssembly as it offers more atomic operations than plain JavaScript, and is based on the code of Rokicki et al. [30]. We use a Python Selenium framework to automatically test and evaluate the power consumption of WebAssembly instructions. Due to the stack machine design of WebAssembly, the output of the previous instruction is the input of the next. Therefore, instructions with different input and output types cannot be called in a row. To address it, we create pairs of complementing operations, *i.e.*, the output type of the first is the input type of the second, and we evaluate them as a whole. In total, we evaluate 211 single and paired instructions.

Web browsers are colossal pieces of software, running computation-heavy tasks: network management, graphical display, cryptographic operations, and client-side operations. This computation can create noise in our measurement, compared to the controlled environment of native power fingerprinting. The design process of the framework is based on lowering as much as possible this noise, while still running the experiments in a standard release browser.

We ran the experiments of this section in Firefox 107 running WebAssembly 1.1. Before starting the actual measurement, the framework loads the attack page in the browser, fetches and instantiates all the tested instructions before starting the measurement. As in the native case, for each instruction, the framework reads the system's total energy consumption, executes the instruction 100 times in the browser, and reads the total energy consumption once again, saving the difference in the power trace. To ensure that the JavaScript components required to run WebAssembly are not creating unwanted execution, we unrolled the loop directly in the WebAssembly script. This allows the most atomic measurement of browser computations and reduces potential noise.

As client-side code runs entirely in a sandbox, it is impossible to use built-in features to measure the power consumption, only to create the artificial power consumption for our experiments. A native component is still needed to read the power consumption. Hence, an attacker sitting in the JavaScript sandbox cannot measure this fingerprint. However, this native component could be integrated into the applicative layer of the browser to provide a strong authentication factor for web browsing. We discuss this limitation further in Sect. 6.2.

5 Results

We evaluate the accuracy of our method by comparing it to the base rate, which is the accuracy of a random guess.

5.1 Classification Pipeline

A trace from our method consists of power and temperature measurements for each instruction. The exact number of samples per trace was not equal among all machine types, as some older microarchitectures do not support all of the instructions in our set. The classification process is as follows:

1. We compute the average temperature for each trace to have a single representative.
2. We exclude outliers using clipping, which is caused by context switches. Specifically, we replace values that are lower than the first percentile of power measurement values with the value of that percentile, and values that are higher than the last percentile with the value of that percentile.
3. We use feature extraction to extract useful information from each trace. The features that we extract include: mean, standard deviation, median absolute deviation, skew, entropy, the value of each percentile between 10 and 90 with jumps of 10, L1 distance between the mean and the median, mean of the sequence of differences, median of the sequence of differences, standard deviation of the sequence of differences, and the number of peaks of the trace.
4. We feed the resulting computed features into a Random Forest classifier.

With the exception of the n_estimators parameter, which is set to 300, we use the default hyper settings for the Random Forest implementation. We used sklearn version 1.0.

5.2 Native-Code Fingerprinting

Classification Using Power Consumption Only. As a first evaluation, we use only the power-consumption features to train the classifier. We use the collection's first 80% of traces as training data and the remaining 20% as testing data for each group. We only use the training set from the initial collection to train the classifier. We balance the datasets by using the same number of traces for each machine. The base rate is 1 divided by the machine count. Figure 2 presents the accuracy of our methods using traces that were gathered the same

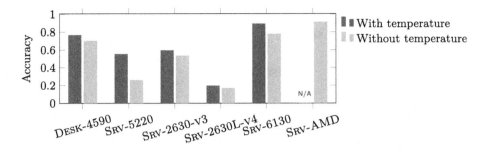

Fig. 2. Summary of classification accuracy results.

day as the training traces. We can see that our method's accuracy is significantly better than base rate for every group of machines. A different classifier is trained for each group of machines. The effect of temperature is explained below.

We also evaluate how our method performs on collections spanning different days. To demonstrate that our method is robust (*i.e.*, above the base rate) over time, we gathered balanced data on various days utilizing DESK-4590 and SRV-5220 group machines. For the DESK-4590 group, the accuracy is $70.14 \pm 0.46\%$ on the test part of the first collection, $67.31 \pm 0.36\%$ on a collection that was done 2 days later and $59.30 \pm 0.49\%$ on a collection that was done 3 days after the first collection, compared to a base rate of 5.88%. For the SRV-5220 group, the accuracy is $25.98 \pm 1.92\%$ on the test part of the first collection, $16.71 \pm 0.50\%$ on a collection that was done 8 days later and $17.14 \pm 0.49\%$ on a collection that was done 10 days after the first collection, compared to a base rate of 0.81%. While the accuracy drops noticeably on days where the classifier was not trained on, the results are still significantly better than the base rate. Since we don't have complete control over SRV-5220 machines due to their location in a shared grid environment, the SRV-5220 dataset has a bigger interval between the training traces and other data collections compared to other groups.

Temperature. To check whether temperature affects our method, we first take a single collection of the SRV-5220 machines, find the median temperature per machine, and split the dataset into 2 parts: traces with temperatures below the median temperature per machine and traces with temperatures above the median temperature per machine. We evaluate the resulting classifiers against a collection that was done 8 days after the training collection, that we also split into colder and hotter traces. We discovered that a classifier that is trained only on the colder (resp. the hotter) traces yields an accuracy of 14.84% (resp. 14.98%) on the test collection, while a classifier that is trained on both hotter and the colder traces yields a higher accuracy of 17.54%. This indicates that the temperature of the CPU while collecting the traces affects our method.

To take temperature into account, we add it as a feature of our classifier on all machines except the SRV-AMD group, which had no operating system support for temperature collection. The process for computing the temperature

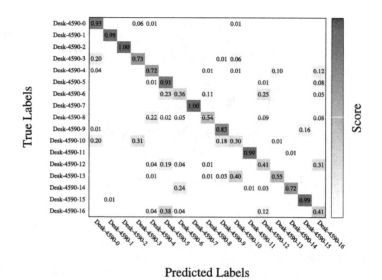

Fig. 3. Confusion matrix for DESK-4590 using all features including temperature.

feature is detailed in Sect. 5.1. As can be seen on Fig. 2, using temperature as a feature improves our method's classification accuracy by a significant margin. Moreover, adding temperature as a feature also makes the system more robust to temporal drift. With the addition of temperature as a feature for the DESK-4590 group, the accuracy is $76.50 \pm 0.67\%$ on the test part of the first collection, $81.51 \pm 0.31\%$ on a collection that was done 2 days later and $70.55 \pm 0.41\%$ on a collection that was done 3 days after the first collection. With the addition of temperature as a feature for the SRV-5220 group, the accuracy is $55.28 \pm 0.62\%$ on the test part of the first collection, $29.75 \pm 0.69\%$ on a collection that was done 8 days later and $28.21 \pm 0.43\%$ on a collection that was done 10 days after the first collection. Figure 3 shows a confusion matrix on DESK-4590 machines when using all features including temperature feature. Rows are the actual machines of these traces, columns are the predicted machines for these traces. We can observe that our model is able to classify machines with high accuracy as we reported earlier. We can also observe that classification errors are not random, but instead tend to form small clusters of machines with similar power consumption.

Classification Using Fewer Instructions. The collection time of a fingerprint, *i.e.*, how long it takes to measure and collect a trace, is an important performance metric. To see how the the data collection time may be reduced, we evaluated each of the instructions in the trace, checking each one's contribution to the classification process. Since the statistical features used as input to our classifier are aggregated from multiple instructions, we could not do this directly, but instead performed an additional analysis: First, we trained a Random Forest classifier on the raw power consumption trace, after clipping outliers but without any additional preprocessing and temperature readings. This classifier has lower

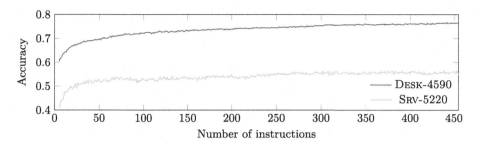

Fig. 4. Effect of number of instructions on accuracy, using a single trace.

accuracy, compared to the classifier that was trained with temperature and features that were extracted using our feature extraction method. However, since this classifier was specifically trained on the power consumption samples, we can use the standard *feature importance score* metric to directly identify those with the highest contribution to accuracy. We ranked the instructions according to their importance, and then used only a subset of the most significant features as input to our statistical feature extraction process. A table containing all evaluated instructions, together with their feature importance score, can be found in the artifact repository.

Figure 4 shows our method's accuracy when using fewer assembly instructions. As shown in the figure, we obtain an accuracy of 63% and 44%, for DESK-4590 and SRV-5220 respectively, by using less than 10 instructions. This is approximately 80% of the peak accuracy obtained using all instructions, which is 75.5% and 55.9% respectively. Even if we use only the 5 most helpful instructions process we obtain a high accuracy of 60.4% and 41.1% respectively. We observe improved accuracy as we use more instructions, up to approximately 300 instructions for DESK-4590 and 250 instructions for SRV-5220. To understand why certain instructions had a higher impact on the classification accuracy than others, we performed a further manual analysis of the most significant instructions, noting the instruction set family of each command, based on the analysis provided by Abel et al. [1]. This annotated version of the instruction table can also be found in the repository. When analyzing the annotated instruction table, we discovered that out of the 20 most helpful instructions, all but one belong to the Advanced Vector Extensions (AVX) and Streaming SIMD Extensions 4 (SSE4) instruction sets. We performed a similar analysis on the WebAssembly dataset, as described in Sect. 4.3, and discovered a similar situation – all but 7 of the 20 most helpful instructions are 128-bit vector instructions. Vectorized instructions are known to use significantly more power compared to regular instructions, probably because they process more data. Our results suggest that this increased power consumption in turn leads to a more distinct power consumption signature, which can be used by our classifier. Interestingly, it is known that the high power consumption of the AVX core requires special handling by the CPU's power monitor, which dynamically powers on the AVX core when

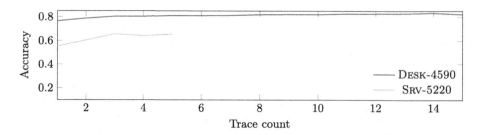

Fig. 5. Effect of number of traces on accuracy, using all instructions.

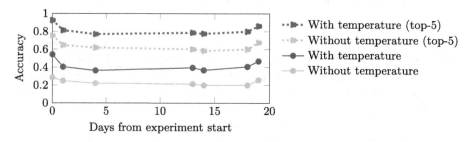

Fig. 6. Effect of temporal drift on accuracy.

these instructions are used. As shown by Schwarz et al. [36], this power-up delay can be used to perform a remote side-channel attack in a different setting.

As mentioned in Sect. 4.2, we did not measure the power consumption of the entire space of valid x86 instructions. Thus, there may be additional instructions which we did not evaluate which have even better performance. In addition, the set of best-performing instructions likely varies between different processor generations and microarchitectures.

Classification Using Multiple Traces. In order to improve the accuracy, at the cost of a longer trace acquisition time, we can gather multiple traces, pass them into the classifier, obtain the probability that each trace corresponds to each class, add the probabilities for each class, and output the class with the largest sum. Figure 5 shows the accuracy as a function of the number of traces used for inference on DESK-4590 machines. It can be seen in the Figure, as we use more traces for a single inference, the accuracy increases until it stabilizes at 11 traces for inference for DESK-4590. We performed a similar evaluation for SRV-5220, using a smaller amount of traces for each prediction, since we collected less traces in this setting. The increase in accuracy when using more traces is also observed for SRV-5220, although the available number of traces per prediction is insufficient to determine when the accuracy reaches a plateau.

Stability of Results Over Time. The ability of fingerprinting methods to fingerprint machines over time is an important measurement to the fingerprint evaluation. To measure the effect of time on our method, we launched a continuous experiment on the SRV-5220 machines, in which we run the collection in

Table 2. Accuracy for multiple cores evaluation on DESK-4590.

Trained on	Tested on		
	Core 1	Core 2	Core 1 & Core 2
Core 1	0.654	0.650	0.652
Core 2	0.621	0.645	0.650
Core 1 & Core 2	0.649	0.651	0.650

best effort mode. In best effort mode we run our collection on a machine as long
the machine is available. In a case that someone orders this machine or that this
machine becomes unavailable, our collection stops until that machine is free or
available again. This experiment can lead to an imbalanced dataset because of
the nature of best effort mode. This collection spans over a period of 19 days and
contains 508634 traces. To evaluate the accuracy of our method over time we
train a classifier on data that originates from the first day of this collection. We
used an equal number of traces per machine to train this classifier. We report
the accuracy of our classifier per day in Fig. 6. We also report the probability
that the correct machine was one of the top 5 outputs of the classifier. Note
that even that our dataset is imbalanced, the classifier was trained on balanced
data and it does not know the imbalanced machine distribution. In this case,
the base rate of our model is a random guess between 105 machines $i.e.$, 0.95%,
since we collected the data in best effort mode and some machines were not
available during the temporal drift data collection. We can observe that our
classification accuracy is better on the day of the collection that the model was
trained on compared to other days. Our model's accuracy drops on traces that
were collected on later days than the collection day of the training traces. We
can observe that our method's accuracy is well above base rate, even for later
collections.

To test an extreme case, we take two collections from 16 machines from the
DESK-4590 group, train a classifier on the first collection and test on the second
collection. The second collection was launched 8 months after the first collection.
This classifier yields an accuracy of 28.50% on the later collection compared to
a base rate of 6.25%, without using the temperature as a feature. While this is
a significant drop in performance, it is still well above the base rate.

Multiple Cores. The behavior of different cores on the same machine is interesting since it can affect our fingerprinting results when core pinning is not
applied. By conducting research on the effect of multiple cores on our fingerprinting method we can conclude whether we fingerprint the core itself or another
hardware component. Data collection is performed using a similar method to
that employed for a single core, except that the pipeline is repeated on different physical cores. The process begins by obtaining a list of all physical cores
on the device, after which the process is pinned to the first core in the list
and all instructions are executed on it. The process is then repeated from the

beginning, this time using the second physical core from the list, resulting in two traces obtained each time. The collected traces were split into two groups, based on the physical cores from which they were obtained. The accuracy of the classifier was evaluated under various core scenarios, as shown in Table 2. The results indicate consistent performance of the classifier across all scenarios, except when trained on traces from core 2 and tested on traces from core 1. In this case, a slight decrease in accuracy was observed. It is worth noting that the accuracy in this evaluation is lower than that reported in Fig. 2, due to the longer trace collection period, which resulted in greater temperature variations compared to the shorter experiment in Fig. 2. To investigate the classifier's ability to distinguish between traces obtained from the same machine but from different cores, a new experiment was conducted, in which the collected traces were grouped by their respective machines. The classifier and pre-processing methods were identical to those used in the previous experiments. For each machine, a classifier was trained to identify whether the trace originated from core 1 or core 2, yielding accuracies ranging from 48.5% to 53.9%, except for one machine that achieved an accuracy of 89.5%. However, all classifiers with accuracies between 48.5% and 53.9% failed to accurately classify the core from which the trace originated, due to the base rate of this experiment being 50%. The feature importance of the classifiers was analyzed to understand why the classifier was successful in identifying the different cores on one particular machine. The maximum feature importance of the classifiers that failed to distinguish between the cores was 0.076, whereas the maximum feature importance of the successful classifier was 0.437, with the 10th percentile of power consumption being the most important feature with a big margin. Upon examining the data from the machine on which the classifier was successful, it was observed that there was a clear separation between the cores based on the 10th percentile of power consumption. This separation is not present in the data from the other machines.

5.3   Portable Fingerprinting

We collected data using our portable method, using the same pre-processing and classifier as the native collections. We do not collect temperature data with this method. This classifier's accuracy is 35.41% for the DESK-4590 machines. This is a significant accuracy drop compared to the native collections due to the high-level nature of the web-based experiment. Web browsers are huge pieces of software, running many other tasks than the WebAssembly instructions, which can result in additional noise, hence power consumption, compared to the controlled environment. Furthermore, we cannot ensure the translation of WebAssembly instructions into native instructions, resulting in less control over the experiment. The results of this PoC are encouraging as they are still well above base rate, showing that power consumption could be used as a strong attribute, yielding a high uniqueness in the browser context. We expect that this accuracy could be improved by adapting the pipeline to browser-specific noise.

6 Discussion

6.1 Related Work

PUFs and PC Fingerprinting. Kohno et al. [17] introduced a technique for remote physical device fingerprinting that is based on clock skew. Using this technique, the authors can determine whether two devices that possibly shifted in time or IP addresses are the same physical device. Sánchez-Rola et al. [32] presented a way to create a fingerprint based on careful analysis of the exact time it takes the device under test to run a fixed benchmark. Their technique was implemented in both native and web-based versions. Unlike of Sánchez-Rola's method, which collects only a single feature at each execution, our method collects multiple data points, each corresponding to the power consumed by a different instruction. We believe that this increases the robustness of our system. Rokicki et al. [29] used WebAssembly instructions as a fingerprinting method, creating a method for detecting hardware processor generation based on the processor's lookahead buffer behavior. Laor et al. [19] showed that it is possible to fingerprint systems based on the individual execution units found inside their Graphical Processing Units (GPUs). Our method explores a similar instance of manufacturing variations, this time inside the CPU itself and not in one of its peripherals. Another way to create a machine fingerprint is by treating the machine hardware as a physically unclonable function, or PUF. Schaller et al. [33] leveraged the Rowhammer attack that flips bits in RAM as a PUF to improve security in commercial, off-the-shelf devices. Suh and Devadas [37] presented a technique that enables low-cost authentication of individual ICs with the use of PUF. Over time, PUF designs have been shown to be vulnerable to machine learning attacks, where the model learns to predict the PUF response after only a few observations [31]. To resist this type of attacks, Vijayakumar and Kundu [41] proposed a novel PUF circuit that, unlike previous work, does not assume the existence of ideal current sources or operating conditions [13,18]. The novel PUF is based on a circuit block and depends on a non-linear voltage transfer function.

PC Power Consumption. Hähnel et al. [9] showed a way to use RAPL-based power consumption to measure and analyze power consumption of individual functions. The authors demonstrated how to use power to characterize the energy costs for decoding video slices. Lipp et al. [23] used power measurement to conduct novel software-based power side-channel attacks on Intel server, desktop, and laptop computers. The authors exploited the unprivileged access to the Intel RAPL interface to leak AES-NI keys from Intel SGX, break kernel address-space layout randomization, infer secret instruction streams and establish a timing-independent covert channel. Von Kistowski et al. [15] noted that PCs have variable power consumption in a performance benchmarking setting. They explored the power consumption of identical CPUs for multiple workloads, and showed that these different CPU samples display statistically significant differences. We extend the work of von Kistowski et al., by turning their observations about power consumption differences into a feature that can be used to tell apart identical devices.

6.2   Limitations

Root Required for Power Measurements. A primary limitation of our scheme is that it only works if the fingerprinting agent can measure power consumption. The easiest way to perform this measurement is through the RAPL interface, which is currently restricted to high-privileged processes. This limits the use of the system to the defensive setting, since there is no way for a malicious fingerprinting attacker (e.g., an intrusive web page) to measure the power consumption of the PC while it is running in user mode.

To address this limitation, we note that there are several works showing how power consumption can be measured indirectly via a side channel on modern PCs. For example, Cohen et al. showed that power consumption can be measured using rowhammer [4], and Wang et al. [42] showed that it is modulated onto the system's clock frequency. Although outside the scope of this work, our method may be turned into an attack by combining our results on fingerprinting with one of these techniques for performing user-land power consumption measurement. On a more cynical note, we observe that features with impact on security are often removed due to security disclosures, but then re-introduced to systems, sometimes with a partial countermeasure, due to external demand for their functionality. For example, high-resolution GPU timers were removed from Chrome 65 after Frigo et al. discovered they can be used for side-channel attacks [6], and then re-enabled with some mitigations in Chrome 70 [28]. Unprivileged access to power measurement, currently disabled due to the work of Lipp et al. [23], may suffer the same fate. In that case, our work will immediately gain an offensive aspect.

Limited Accuracy and Stability. Our method has limited accuracy and stability over time. In particular, it is unable to identify a single computer from a large population with sufficient accuracy to be used as a single source of authentication. While this accuracy may be improved with a better choice of instruction mix and a more refined machine learning pipeline, the limitation ultimately stems from the fact that power consumption is a physical property which does not depend on the workload alone, but also on external influences both inside and outside the device under test, such as temperature, incoming noise on the system's power supply, and even activity of other computers on the same power distribution network [43]. Our method is therefore the most useful when integrated as a contributing feature into an existing fingerprinting system, or when used as a first line of defense before resorting to more intrusive fingerprinting methods or even asking the user to manually authenticate [2].

No Evaluation in the Wild. This work only evaluates the effectiveness of our fingerprinting method in a lab setting, when telling apart identical computers. It would be interesting to consider the ability of power consumption measurements to tell apart computers with diverse hardware and software configurations in the wild. We note that, in practice, a fingerprinting scheme would make use of all information available in the system, including deterministic metrics such as the list of installed hardware and software, the network address, the time zone, and so on. Thus, the power fingerprint would actually be used in a setting very similar to the lab setup, to identify the computer among a small cluster of candidates with identical configurations. As Laor et al. observed, this setting actually improves the performance of non-deterministic fingerprinting methods [19].

Slow Data Collection. Our fingerprinting agent takes about 7 s to collect a full power trace of 455 instructions, from the system. While this may be appropriate in some settings, speeding up the process will definitely make it more practical. One of the main reasons for this long runtime is the design of the agent, which is built for reproducibility rather than performance. We analyzed the runtime of the code and found that the actual measurements account for less than 25% of its runtime, with the rest dedicated to logging, data management scripts and diagnostic printouts. A practical solution written in a high-performance language could avoid these extra steps. Going even further, as noted in Sect. 5.2, even very small number of instructions is enough to capture more than 80% of the system's peak accuracy. In particular, a performance-oriented fingerprinting scheme can obtain usable results after profiling no more than six instructions.

6.3 Countermeasures

Even though there is no immediate offensive application for our work, it is still interesting to consider how a system can remain unidentifiable, even in the presence of a power-based fingerprinting agent. The most straightforward approach to avoiding fingerprinting would be introducing noise to the power consumption measurement. This can be internal noise, generated by executing code on the DUT, or external noise, generated by plugging in a noisy device, such as a microwave oven, to the same power distribution line as the PC and running it when the fingerprint is collected. We note that this mechanism only decreases the signal-to-noise ratio of the system, requiring more traces to be collected for a reliable reading, but only partially eliminates the fingerprinting capability.

Another interesting, but unfortunately ineffective, countermeasure would be to use the *power capping* mechanism available in modern processors. This mechanism places a hard limit on the total power consumption of the device under test by dynamically controlling its clock. Obviously, if the power cap is set aggressively, all of the instructions executed on the machine will have the same power consumption, reducing the accuracy of our method. unfortunately, as recently observed by Liu et al. [24], fixing the power consumption only moves the side-channel information into the frequency domain, with higher-power instructions simply taking longer to execute than lower-power instructions. Since our fingerprinting agent already logs the time taken to execute each instruction, it will be able to overcome this countermeasure.

7   Conclusions

As a result of identity theft and authentication attacks, device identification has become an important topic in recent years. However, most fingerprint methods such as [19, 30] rely on the differences between machines with different software or hardware. Consequently, these kind of fingerprints would not be able to distinguish between identical devices in the same environment. In this paper, we created a new method to identify devices with the same hardware and software characteristics, based only on the CPU x86-64 micro-architectural properties. We used the power consumption similar to PUF concept as the main feature to create a fingerprint that can tell apart identical devices in a way that other fingerprints cannot separate. Our method shows a way to use power as a foundation for a robust fingerprint, as the power consumption of devices varies slightly. Moreover, we showed that with the use of other CPU properties, such as CPU temperature, as another feature, the fingerprint is more accurate, robust, and stable. Through comprehensive evaluation, we showed that our technique can distinguish between identical sets of machines with different micro-architectures (Intel, AMD) and can be used not only on endpoint machines but also on servers. Furthermore, we showed that this technique can be executed natively by using the x86-64 instruction set, and portable by using the WebAssembly instruction set.

Future Work. Our work lays the foundation for future work on authentication methods based on micro-architectural features. In terms of future work, we first note our work requires ring 0 access as there is no other way, to our knowledge, to accurately measure power consumption from software. Once it becomes possible to measure power consumption with ring 3 privileges, our technique can be used as an authentication method to fingerprint data, both natively and portably. Another future direction would be an in-the-wild evaluation on a machine set larger than 130 devices. While most of our measurements were performed on a single core, our results indicate that there may be some added value from extracting fingerprints from multiple cores. It would be interesting to find the optimal combination of instructions, cores and sample counts that can obtain the best accuracy for a given sampling time budget. Another direction is performance improvement. In this work, we focused on the system's reproducibility and readability, rather than the data collection time. The speed of the data gathering can be increased by several methods, such as moving from a scripting harness to 100% native code, reducing the instruction set as shown in Fig. 4, or reducing the number of iterations for each instruction. Finally, we showed that CPU temperature can increase the accuracy rate and can be used as another feature in the classification process. As temperature and power are not the only CPU micro-architectural properties, we infer that more features can be used in the identification, which will improve its robustness, accuracy, and stability. Furthermore, we believe that a power consumption feature can be added to existing authentication methods [7, 19, 30, 38, 40].

Acknowledgments. Code to collect power consumption traces is based on Gras et al. [8]. This work has been partly funded by the ANR-19-CE39-0007 MIAOUS. Experiments presented in this paper were carried out using the Grid'5000 testbed, supported by a scientific interest group hosted by Inria and including CNRS, RENATER and several Universities as well as other organizations (see https://www.grid5000.fr).

Artifact Availability.. Our developed code and data artifacts are available at https://github.com/FingerInThePower/Finger_In_The_Power, including code for power consumption trace collection for each of the architectures used as well as the portable code, our datasets used for the results section with the results, and the machine learning pipeline with the pre-processing procedures.

References

1. Abel, A., Reineke, J.: uops.info: Characterizing latency, throughput, and port usage of instructions on intel microarchitectures. In: ASPLOS (2019)
2. Alaca, F., van Oorschot, P.C.: Device fingerprinting for augmenting web authentication: classification and analysis of methods. In: ACSAC, pp. 289–301 (2016)
3. Cherkaoui, A., Bossuet, L., Seitz, L., Selander, G., Borgaonkar, R.: New paradigms for access control in constrained environments. In: ReCoSoC. IEEE (2014)
4. Cohen, Y., et al.: Hammerscope: observing DRAM power consumption using rowhammer. In: CCS (2022)
5. Colombier, B., Bossuet, L.: Survey of hardware protection of design data for integrated circuits and intellectual properties. IET Comput. Digit. Tech. **8**(6), 274–287 (2014)
6. Frigo, P., Giuffrida, C., Bos, H., Razavi, K.: Grand pwning unit: accelerating microarchitectural attacks with the GPU. In: S&P (2018)
7. van Goethem, T., Scheepers, W., Preuveneers, D., Joosen, W.: Accelerometer-based device fingerprinting for multi-factor mobile authentication. In: 8th International Symposium on Engineering Secure Software and Systems (ESSoS) (2016)
8. Gras, B., Giuffrida, C., Kurth, M., Bos, H., Razavi, K.: Absynthe: automatic blackbox side-channel synthesis on commodity microarchitectures. In: NDSS (2020)
9. Hähnel, M., Döbel, B., Völp, M., Härtig, H.: Measuring energy consumption for short code paths using RAPL. SIGMETRICS Perform. Eval. Rev. **40**(3), 13–17 (2012)
10. Holcomb, D.E., Burleson, W.P., Fu, K.: Power-up SRAM state as an identifying fingerprint and source of true random numbers. IEEE Trans. Comput. **58**(9), 1198–1210 (2009)
11. Hupperich, T., Hosseini, H., Holz, T.: Leveraging sensor fingerprinting for mobile device authentication. In: DIMVA (2016)
12. Intel: Running Average Power Limit Energy Reporting/INTEL-SA-00389. https://www.intel.com/content/www/us/en/developer/articles/technical/software-security-guidance/advisory-guidance/running-average-power-limit-energy-reporting.html (2022)
13. Kalyanaraman, M., Orshansky, M.: Novel strong PUF based on nonlinearity of MOSFET subthreshold operation. In: HOST (2013)
14. Khan, K.N., Hirki, M., Niemi, T., Nurminen, J.K., Ou, Z.: RAPL in action: experiences in using RAPL for power measurements. ACM Trans. Model. Perform. Evaluation Comput. Syst. **3**(2), 9:1–9:26 (2018)
15. von Kistowski, J., Block, H., Beckett, J., Spradling, C., Lange, K., Kounev, S.: Variations in CPU power consumption. In: ICPE, pp. 147–158. ACM (2016)
16. Kocher, P., Jaffe, J., Jun, B.: Differential power analysis. In: Wiener, M. (ed.) CRYPTO 1999. LNCS, vol. 1666, pp. 388–397. Springer, Heidelberg (1999). https://doi.org/10.1007/3-540-48405-1_25
17. Kohno, T., Broido, A., Claffy, K.C.: Remote physical device fingerprinting. In: S&P (2005)

18. Kumar, R., Burleson, W.P.: On design of a highly secure PUF based on non-linear current mirrors. In: HOST, pp. 38–43. IEEE Computer Society (2014)
19. Laor, T., et al.: DrawnApart: a device identification technique based on remote GPU fingerprinting. In: NDSS (2022)
20. Laperdrix, P., Avoine, G., Baudry, B., Nikiforakis, N.: Morellian analysis for browsers: making web authentication stronger with canvas fingerprinting. In: DIMVA (2019)
21. Laperdrix, P., Rudametkin, W., Baudry, B.: Beauty and the beast: diverting modern web browsers to build unique browser fingerprints. In: S&P (2016)
22. Lipp, M., Gruss, D., Schwarz, M.: AMD prefetch attacks through power and time. In: USENIX Security Symposium (2022)
23. Lipp, M., et al.: PLATYPUS: software-based power side-channel attacks on x86. In: S&P (2021)
24. Liu, C., Chakraborty, A., Chawla, N., Roggel, N.: Frequency throttling side-channel attack. In: CCS (2022)
25. Mangard, S., Oswald, E., Popp, T.: Power Analysis Attacks. Springer, Boston, MA (2007). https://doi.org/10.1007/978-0-387-38162-6
26. Marchand, C., Bossuet, L., Mureddu, U., Bochard, N., Cherkaoui, A., Fischer, V.: Implementation and characterization of a physical unclonable function for IoT: a case study with the TERO-PUF. IEEE Trans. Comput. Aided Des. Integr. Circuits Syst. **37**(1), 97–109 (2018)
27. Messerges, T.S., Dabbish, E.A., Sloan, R.H.: Power analysis attacks of modular exponentiation in smartcards. In: CHES (1999)
28. Moenig, M.: Webgl2: EXT_disjoint_timer_query_webgl2 failing in beta of 65 (2018). https://bugs.chromium.org/p/chromium/issues/detail?id=820891
29. Rokicki, T., Maurice, C., Schwarz, M.: CPU port contention without SMT. In: ESORICS (2022)
30. Rokicki, T., Maurice, C., Botvinnik, M., Oren, Y.: Port contention goes portable: port contention side channels in web browsers. In: ASIACCS (2022)
31. Ruhrmair, U., Solter, J.: PUF modeling attacks: an introduction and overview. https://doi.org/10.7873/DATE2014.361
32. Sánchez-Rola, I., Santos, I., Balzarotti, D.: Clock around the clock: time-based device fingerprinting. In: CCS (2018)
33. Schaller, A., et al.: Intrinsic rowhammer pufs: leveraging the rowhammer effect for improved security. CoRR abs/1902.04444 (2019)
34. Schaller, A., et al.: Decay-based DRAM pufs in commodity devices. IEEE Trans. Dependable Secur. Comput. **16**(3), 462–475 (2019)
35. Schellenberg, F., Gnad, D.R.E., Moradi, A., Tahoori, M.B.: An inside job: remote power analysis attacks on FPGAs. In: DATE (2018)
36. Schwarz, M., Schwarzl, M., Lipp, M., Masters, J., Gruss, D.: Netspectre: read arbitrary memory over network. In: ESORICS (2019)
37. Suh, G.E., Devadas, S.: Physical unclonable functions for device authentication and secret key generation. In: DAC, pp. 9–14. IEEE (2007)
38. Tehranipoor, F., Karimian, N., Yan, W., Chandy, J.A.: Dram-based intrinsic physically unclonable functions for system-level security and authentication. IEEE Trans. Very Large Scale Integr. Syst. **25**(3), 1085–1097 (2017)
39. Tian, S., Xiong, W., Giechaskiel, I., Rasmussen, K., Szefer, J.: Fingerprinting cloud FPGA infrastructures. In: FPGA (2020)
40. Trampert, L., Rossow, C., Schwarz, M.: Browser-based CPU fingerprinting. In: ESORICS (2022)

41. Vijayakumar, A., Kundu, S.: A novel modeling attack resistant PUF design based on non-linear voltage transfer characteristics. In: DATE, pp. 653–658. ACM (2015)
42. Wang, Y., Paccagnella, R., He, E.T., Shacham, H., Fletcher, C.W., Kohlbrenner, D.: Hertzbleed: turning power side-channel attacks into remote timing attacks on x86. In: USENIX Security Symposium (2022)
43. Yang, L.,et al.: Remote attacks on speech recognition systems using sound from power supply. In: USENIX Security Symposium (2023)

PWRLEAK: Exploiting Power Reporting Interface for Side-Channel Attacks on AMD SEV

Wubing Wang[1], Mengyuan Li[1], Yinqian Zhang[2], and Zhiqiang Lin[1(✉)]

[1] The Ohio State University, Columbus, OH 43210, USA
{wang.11488,li.5733}@osu.edu, yinqianz@acm.org
[2] Southern University of Science and Technology, Shenzhen 518055,
Guangdong, China
lin.3021@osu.edu

Abstract. An increasing number of Trusted Execution Environment (TEE) is adopting to a variety of commercial products for protecting data security on the cloud. However, TEEs are still exposed to various side-channel vulnerabilities, such as execution order-based, timing-based, and power-based vulnerabilities. While recent hardware is applying various techniques to mitigate order-based and timing-based side-channel vulnerabilities, power-based side-channel attacks remain a concern of hardware security, especially for the confidential computing settings where the server machines are beyond the control of cloud users. In this paper, we present PWRLEAK, an attack framework that exploits AMD's power reporting interfaces to build power side-channel attacks against AMD Secure Encrypted Virtualization (SEV)-protected VM. We design and implement the attack framework with three general steps: (1) identify the instruction running inside AMD SEV, (2) apply a power interpolator to amplify power consumption, including an emulation-based interpolator for analyzing purposes and a more general interrupt-based interpolator, and (3) infer secrets with various analysis approaches. A case study of using the emulation-based interpolator to infer the whole JPEG images processed by libjpeg demonstrates its ability to help analyze power consumption inside SEV VM. Our end-to-end attacks against Intel's Integrated Performance Primitives (Intel IPP) library indicates that PWRLEAK can be exploited to infer RSA private keys with over 80% accuracy using the interrupt-based interpolator.

1 Introduction

Private data is becoming an important asset for our society and personal life. More and more data-driven applications and technologies are introduced to release the value of data to the greatest extent, which, however, can potentially put personal or technical data at risk of leakage. Such concerns are especially perceptible in the cloud computing domain, where numerous cloud tenants rent cloud services from cloud service providers, upload and process sensitive data on the cloud. Potential safety hazards can occur in two ways: (1) malicious cloud tenants steal data from other cloud tenants, and (2) a curious or malicious cloud service provider monitors or steals data from its cloud

D. Gruss et al. (Eds.): DIMVA 2023, LNCS 13959, pp. 46–66, 2023.
https://doi.org/10.1007/978-3-031-35504-2_3

tenants directly. Therefore, there is an urgent demand for the industries to come up with techniques that can guarantee data security in the cloud computing environment.

Trusted Execution Environment (TEE) is one of those techniques that meets the above requirement and has already been adopted in mainstream cloud service providers, such as Google cloud [17], Microsoft Azure [36], and Amazon AWS [1]. With the help of trustworthy hardware (*e.g.*, CPU and memory encryption engine), TEE provides hardware-guaranteed isolation to protect cloud user's data from other cloud users or even the cloud service provider. When TEEs are enabled, even the highest privilege software (*e.g.* operating system, hypervisor, *etc.*) cannot directly access cloud user's data. With such promising security guarantee, CPU vendors, such as Intel, AMD or ARM, all released server-level processors that support TEE features to protect VMs running on the cloud, including the current available AMD Secure Encrypted Virtualization (SEV) [2], and the upcoming Intel Trust Domain Extensions (TDX) [4] and ARM Confidential Computing Architecture (ARM CCA) [3].

However, recent research [23,26,30] showed that different types of side-channel attacks could be exploited to steal TEE-protected secrets. Among different types of side-channel attacks, the power side-channel attack plays a very important role, where the untrusted cloud service provider can easily collect the power consumption of TEE and steal secrets. Platypus [30] first examines the threat of software-based power-based side-channel attacks in cloud-based TEEs. Using power consumption reporting interfaces, Platypus successfully demonstrates that attackers can obtain power consumption with instruction-level granularity through APIC interrupts [44], and can use fine-grained power data to carry out a series of end-to-end attacks, including breaking KASLR and breaking constant-time cryptographic implementations (AES-NI) used by Intel SGX. The feasibility of power-based side-channel attacks in SEV was also discussed in [30].

Inspired by previous work, in this paper, we aim to explore the power-based side-channel attacks in the AMD SEV environment. We introduce PWRLEAK, a software-based power side-channel framework that analyzes power consumption in AMD SEV-only VMs (excluding the newer SEV-ES [21] and SEV-SNP [6] versions). Specifically, PWRLEAK uses the AMD power-reporting features to monitor the program execution inside AMD SEV VMs, and then infers secrets from the VMs. We first analyze the possibility of using power information to distinguish instructions in SEV. We show that different instructions have different power consumption, and the same instruction with various operands also has distinguishable power differences. Based on such observation, PWRLEAK makes use of page-table-based controlled channel [23] to intercept VM's execution in real-time and then infers the secret inside the SEV VM by inferring executed instructions and their operands based on distinguishable power consumption.

We test two interpolators that could amplify and produce distinguishable power consumption of a single instruction on the AMD platform: an emulation-based interpolator and an interrupt-based interpolator. The emulation-based interpolator can acquire higher-resolution power information by emulating the execution of instructions, which acts as an ideal tool to compare power consumption for analysis purposes. The more general interrupt-based interpolator ports an existing APIC-based amplifier introduced by Platypus [30] to AMD platform and can amplify the power information with interrupts by forcing the re-execution of instructions. To demonstrate the capability of PWRLEAK, we showed that the emulation-based interpolator can be used to analyze power consumption inside SEV VM and recover JPEG images processed by libjpeg library.

We further demonstrated that PWRLEAK could steal RSA private keys from the Intel IPP library with over 80% accuracy using the interrupt-based interpolator. To the best of our knowledge, PWRLEAK is the first power-based side-channel attack that extracts secrets from AMD SEV-protected VMs. The prototype of PWRLEAK has been made public available at github.com/OSUSecLab/PWRLEAK. The contributions of the paper are summarized as follows:

- **An instruction-level power consumption study inside SEV VM.** We measure the power-based information leakage towards AMD SEV VMs, and figure out that the power information can be used to differentiate instructions and their operands running inside AMD SEV VMs.
- **Test power interpolators on AMD SEV.** We test two power interpolators that take advantage of the instruction emulation function or the advanced programmable interrupt controller (APIC) to amplify the power consumption of a single instruction. We show that these two interpolators are useful to amplify and analyze the energy consumption of executed instructions in SEV VMs.
- **A new attack on AMD SEV VM.** We also propose PWRLEAK, a new power attack framework on AMD SEV VMs, and successfully steal secrets from a VM protected by the baseline SEV version. The feasibility and limitations of similar attacks but in newer versions of SEV (SEV-ES and SEV-SNP) and the corresponding countermeasures are also discussed in the paper.

Responsible Disclosure. We disclosed the proposed findings and attacks to AMD in April 2023. At the time of writing, AMD has acknowledged our findings and provided a tracking ID for future communications. However, as discussed in Sect. 6, neither attack method presented in this paper could be directly conducted against the newer versions of SEV (*e.g.*, SEV-ES and SEV-SNP). The emulation-based interpolator acts as an analysis tool and is not expected to work for SEV-ES or SEV-SNP. For the interrupt-based interpolator, our paper makes use of it to demonstrate that power-based side-channel attacks can work in SEV, but we did not conduct relevant experiments in SEV-ES or SEV-SNP. We can foresee that there may be a lot of additional noise caused by additional protections enabled by SEV-ES and SEV-SNP, such as register encryption or ownership check, which prevents PWRLEAK from working directly. Therefore, VM protected by SEV-ES or SEV-SNP will not be affected by PWRLEAK.

2 Background

2.1 AMD Secure Encrypted Virtualization (SEV)

AMD first introduced Secure Encrypted Virtualization (SEV) in 2016 [2], which is a hardware-based technology designed to protect virtual machines (VMs) against both privileged software attackers and physical attackers on a remote platform. To protect the confidentiality of guest VMs' code and data, SEV provides necessary isolations for data (*e.g.*, cache and TLB) within the CPU chip, and encrypts VM's memory using memory encryption [48]. AMD later introduced SEV-ES (Encrypted State, the second generation of SEV [21]) to add additional protection towards VM's unencrypted register states during VM-hypervisor world switch. Lately, in order to add additional memory integrity protection and defend against several controlled-channel attacks (page table manipulation attacks [47]), AMD introduced the third generation of SEV on Zen 3 architecture, called SEV Secure Nested Paging (SEV-SNP [6]). Due to the strong security guarantee and user-friendly mode provided, AMD SEV has already been adopted by some public cloud service providers, including Google Cloud [17] and Microsoft Azure [36].

2.2 Hardware Power Reporting Feature

The hardware power reporting interfaces provided by commodity processor vendors, such as Intel and AMD, allow software to monitor and control CPU's power consumption. The reporting interfaces related to power consumption specified in the AMD manual [52] include each CPU core's effective frequency and power consumption:

The Effective Frequency, which monitors the real CPU frequency of each core with the Max Performance Frequency Clock Count (MPERF) and Actual Performance Frequency Clock Count (APERF) `rdmsr` MSR registers. These two registers can be accessed in kernel mode using `rdmsr` and `wrmsr` instructions. Users can calculate the effective frequency of a core over a software-determined window of time.

Processor Core Power Consumption, which provides power consumption for a given core over a software-determined time interval in `MSR_CORE_ENERGY_STAT` MSR. The value of the register is the cumulative energy consumption of a given CPU core. The sampling interval of `MSR_CORE_ENERGY_STAT` is 1 *ms*. Compared to the power interfaces in Intel Processors, which have a 50 μs sampling interval, AMD processors sample power consumption in a much coarser granularity.

2.3 Power-Based Side-Channel Attacks

Power-based side-channel attacks exploit the collected power information to distinguish victim's behaviors or infer secret from the victim. The power-based side-channel attacks can be further classified as *hardware-based power attacks* and *software-based power attacks*.

Hardware-Based Power Side-Channel Attacks. The *hardware-based power attacks* can usually acquire power consumption data with a higher granularity using an independent device. Existing attacks showed that power consumption data collected in this way could help an attacker identify the executed instruction [41,42,45] or infer the execution trace of programs [16]. Lately, researchers showed that hardware-based power side-channel attacks could successfully steal the RSA private key algorithm [20,53] or AES private keys [13,37,40].

Software-Based Power Side-Channel Attacks. Software-based power side-channel attacks rely on software-based power reporting interfaces to collect power consumption data. There are many efforts to explore software-based power side channels in smartphones [12,35], which could fingerprint application being used or identify user's movement. Recent work in the past few years has used software-based power side channels

50 W. Wang et al.

to break isolation protections provided by modern desktop or server processors. The two most relevant papers related to this article include Platypus [30], which focuses on attacking Intel SGX, and another software-based power side-channel attack [29], which focuses on AMD CPUs. On Intel CPUs, Platypus attacks [30] showed for the first time that attackers could distinguish different executed instructions and their operands by collecting power consumption information from the Intel Running Average Power Limit (RAPL) interface. With the help of APIC-timer interrupt, the attacker could get execution control with instruction-level granularity. These side-channel information could later be used to leak secret keys from the constant-time AES-NI implementation used by Intel SGX, break KASLR, and establish a time-independent convert channel. The power consumption of different instructions, but in AMD 's Zen microarchitecture, and the feasibility of power-based side-channel attacks in SEV was also studied or discussed in the paper. Lipp *et al.* [29] later demonstrated the danger of power-based side-channel attacks in modern AMD processors through several end-to-end attacks, which successfully broke KASLR, stole kernel secrets and established a covert channel from unprivileged attackers. In their attacks, they combined power consumption with prefetch to infer system states and steal secrets. Inspired by those existing papers, in our paper, we focus on AMD's Zen microarchitecture and explore the feasibility of stealing secrets from AMD SEV-protected VMs using software-based power side-channel attacks.

2.4 Common Power Analysis Methods

Simple Power Analysis method and *Cross Correlation Analysis* method are two methods widely used in power-based side channel attacks. Simple Power Analysis (SPA) is a technique that differentiates various operations by distinguishing individual power patterns [51]. Based on the method used to recognize and distinguish power patterns, the SPA attack can be further categorized into two types: the visual SPA attack [32] and the template-based SPA attack [11]. The visual SPA attack manually inspects and recognizes the difference in the power traces, and the template-based SPA attack uses the extracted mathematical statistic template to analyze the power traces. Cross Correlation Analysis (CCA) [33] uses the correlation coefficient to measure power traces to differentiate two inputs. Specifically, if two inputs are similar, the correlation coefficient value is high; otherwise, the correlation coefficient value is low.

Power Consumption Sampling with Instruction-Level Granularity. To achieve a sampling rate with instruction-level granularity, previous attacks [26,30] usually utilized APIC timer interrupts to allow the target to execute a single or multiple instructions before being halted. On Intel platforms, SGX-STEP [44] first showed that an attacker could use APIC timer interrupts to execute zero-step or single-step SGX enclaves with instruction-level granularity. Platypus [30] then first combined this timer interrupt-based technique together with the power consumption interface to reveal the relationship between power consumption and different instructions. A similar methodology was also studied on the AMD platform to monitor the states of SEV VMs. CipherLeaks attack [26] used the APIC timer interrupt to step AMD SEV VM's execution inside an instruction page. In this paper, we collect the power consumption at the instruction level by adopting the same APIC-based sampling method presented in Platypus [30].

3   Exploring Power Consumption Leakage

This section consists of several experiments that aim to exploit the ability of hardware power reporting interfaces in AMD platform and collect the ground truth of the relationship between the power consumption with behaviors of a SEV-protected VM. All the experiments were conducted on a blade server with an 8-Core AMD EPYC 7251 Processor. The host OS runs Ubuntu 64-bit 18.04 with kernel version 4.20.0. The guest VMs run the same kernel version and are configured with 4 virtual CPUs, 4 GB memory, and 30 GB local disk as SEV official GitHub repository suggested [8].

3.1   Synchronous Power Measurement

To accurately measure power consumption, we first introduce a synchronous power measurement method to collect the ground truth and explore the relationship between instructions and the corresponding power consumption. More specifically, we first modified the `CPUID` handler in the KVM to act as an indicator of the start and end of a power trace, so that we could accurately locate the measured behaviors inside the VM. Then, for each instruction to be tested, we used two `CPUID` instructions to indicate the start and end points. We executed each instruction $100,000$ times inside the VM, and measured the overall power consumption on the CPU core. By dividing the total power consumption by $100,000$, we could know the power consumption of that instruction. Although the power consumption of `CPUID` is also included in the results, it is negligible compared to the power consumption caused by $100,000$ repeated instructions.

3.2   Instruction Power Consumption

Our experiment results showed that even under the protection of AMD SEV, different instructions, different operands, and different loaded data all produce distinguishable power consumption-based side-channel information.

Distinguishing Instructions. With the synchronous power measurement approach, we measure the power consumption of the instructions $100,000$ times and calculate the median power consumption of each instruction. We choose instructions that are commonly used and related to cryptography for demonstration. The data are reported in

Table 1. Energy consumption of instructions.

Instruction	Core Power $(10^4 mW)$	Instruction	Core Power $(10^4 mW)$
aesdec xmm1, xmm2	4.266	inc r64	1.488
aesdeclast xmm1, xmm2	4.166	mov mem, r64	1.700
aesenc xmm1, xmm2	5.340	mov r64, r64	0.387
aesenclast xmm1, xmm2	5.251	clflush mem	69.600
aesimc xmm1, xmm2	1.123	xor r64, r64	1.444
pclmullqlqdq xmm1, xmm2	7.150	fscale	40.599
dec r64	1.511	nop	0.554
imul r64, r64	4.633	rdrand r64	29.540

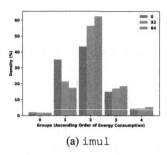
(a) imul

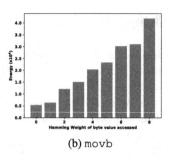
(b) movb

Fig. 1. Power difference because of operand's hamming weights.

Table 1. There are four columns in this table; the first and the third column are the instructions been tested, and the second and the fourth column are the core power consumption for the given instruction. For instance, running instruction aesdec 100, 000 times consumes 4.266×10^4 milliwatts. The differences in power consumption among the various instructions are noticeable. For instance, the aesdec instruction between two registers has 4.26×10^4 core power consumption, while the aesenc instruction between two registers has 5.34×10^4 core power consumption.

Distinguishing Operands. The power information can also be used to differentiate operands of the same instruction. Furthermore, we collected and explored the power difference caused by different operands of the same instruction. Our evaluation suggests that the exact value of the operands is hard to be distinguished. However, operands with different Hamming weights can be differentiated. We use the imul instruction to demonstrate this result. For the operand (64-bit), we selected operands with Hamming weights of 0, 32, and 64 bits. We measured each operand 100, 000 times and got the maximum and minimum value of energy consumption. After dividing the energy interval into five even groups, we counted the number of results for each interval. As shown in Fig. 1(a), the x-axis is the group number, with the energy consumption of each group sorted by ascending order; the y-axis is the percentage of the result that each group contains. In summary, the energy difference in the operand with various Hamming weights is observable; the operand with a lower Hamming weight has relatively less energy consumption.

Distinguishing Loaded Data. Similarly, we measured the relationship between the data loaded from the cache and their power consumption. We used the movb instruction to demonstrate this experiment, which could read one byte of data from the cache line. We generated 256 different pieces of data, which cover all possible values for a byte, and categorized them into nine different groups based on the Hamming weight (*e.g.* 0, 1, 2, ..., 8). After measuring the power consumption that loads every data 100, 000 times, we calculated the average power consumption of the data in each of the nine groups and presented the result in Fig. 1(b). The x-axis is the Hamming weight of each group, and the y-axis is the average power consumption of each group. The results suggest that loading the data with larger Hamming weights could consume more energy, which is distinguishable.

4   PWRLEAK Design

In this section, we present PWRLEAK, an attack framework to differentiate power consumption caused by instructions running inside SEV-protected VMs, and steal secrets from the victim VM.

4.1   Threat Model

In this paper, we consider the same threat model as AMD SEV's threat model [2], where the adversary is a privileged software attacker who does not know the data protected by the SEV-enabled guest VM, and cannot control any program running inside the guest VM. We further assume that the adversary has the pre-knowledge of the target program's binary running inside the VM (*e.g.*, a specific cryptography library), including detailed information such as control flow and function calls of the target program.

4.2   Overview of PWRLEAK

Even though our synchronous power measurement (discussed in Sect. 3) suggests that different behaviors inside SEV VMs lead to different power consumption. There are two main challenges left for a real world power-based side-channel attack. First, the synchronous power measurement approach used to measure repeated instructions is not practical. Second, the low power consumption sampling rate (1 ms) in AMD processors also limits the practicality of a real attack. To overcome these two challenges, we introduce PWRLEAK, whose general components are shown in Fig. 2. *Instruction Identification* is a component used to locate some target instructions in a specific program running inside VMs. *Power Interpolator* is a component that can amplify the power consumption of target instructions, so that the amplified power data is sufficient for PWRLEAK to distinguish different instructions via AMD's coarse-grain hardware power reporting interfaces. *Power Attack* is a component that can run offline, possibly on a separate machine, and infers secrets by analyzing the collected power data.

Instruction Identification. To locate a specific instruction of the program running inside AMD SEV VMs, we use both the page-level memory access pattern [50] and the APIC single-step [44]. We first locate the page of the target instruction with the page-level memory access pattern, then use APIC single-step to further locate the target instruction inside this page. Need to note that the attacker can directly check the rip

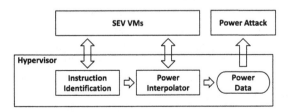

Fig. 2. PWRLEAK Overview.

register to get the number of executed instructions. For a newer version such as SEV-ES or SEV-SNP, the attacker may need to check the ciphertext of the `rip` register to distinguish a single-step from a zero-step.

Power Interpolator. After locating the instruction, we measure the power consumption of these instructions by monitoring the core energy consumption MSR register (`MSR_CORE_ENERGY_STAT`). As the register is updated in a related low rate (*e.g.*, $1ms$), it is hard to measure the power consumption of a small gadget of instructions (*i.e.* one instruction). To solve this problem, in Sect. 4.4, we test two power interpolators that can be used to amplify the power consumption of a single instruction.

Power Attack. Finally, the attackers conduct a power attack by analyzing the power consumption. As we assume that the attacker has knowledge about the program binary to be attacked, the secret about this program can be inferred by distinguishing different operations in critical locations.

4.3 Instruction Identification

To perform the attack, the adversary first needs to pinpoint a specific location (*e.g.*, an instruction inside its instruction page) in the program. In this work, we use the page sequence matching to pinpoint an instruction page, and use the SEV VM single-step to step to an instruction.

Page Sequence Matching. Previous works [50] have proved that the page sequence caused by page faults can be used by the adversary to successfully locate a certain instruction page of a target program. Similarly, in the SEV environment, the adversary collects the page fault sequence for the known target binary and uses this sequence to identify the target instruction page.

To trigger and monitor the page-level access pattern, we first clear the *present* bit in the page table entry for all pages mapped to the VM. Then we monitor the VMEXIT event (*e.g.*, the `handle_exit` function in the kernel). When the VMEXIT is triggered by a page fault, we collect the corresponding page address. Finally, PWRLEAK identifies the specific instruction page using the pre-collected page access pattern.

Unlike the traditional page table walk, SEV adopts a nested page table to maintain the address transmission between the guest virtual address (GVA) and the host physical address (HPA). Specifically, the nested page table consists of a guest page table and a host page table. The guest page table maintains the mapping between GVA and the guest physical address (GPA), and is within the protection of SEV. The host page table maintains the mapping between GPA and HPA and is under the hypervisor's control. Thus, the privileged adversary knows the mapping between GPA and HPA, but will not directly know the relationship between GVA and GPA. Therefore, instead of directly using the GPA to construct the page access pattern, PWRLEAK uses the address interval between two consecutive pages. For instance, we let p_i be the GPA of the i_{th} page access. Then, the page access pattern can be presented as the following: $S = \{p_1 - p_0, p_2 - p_1, \cdots, p_i - p_{i-1}, \cdots\}$.

SEV VM Single-Step. To further improve the attack, it is necessary to narrow down the granularity of the attack to several instructions. PWRLEAK uses the APIC-based

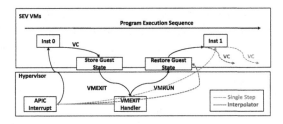

Fig. 3. APIC for the single-step and interpolator. (Color figure online)

single-step to identify the instruction in which we are interested. Similarly to the method introduced in Platypus [30], SGX-Step [44] and Cipherleaks [26], we use the APIC interrupt to force the VM to VMEXIT after a single instruction. By carefully setting the APIC interrupt's timer, the program running inside the SEV VM can be single-stepped.

In particular, as shown in Fig. 3, the first APIC interrupt arrives when the SEV VM is executing instruction 0, which raises a VMEXIT after instruction 0 is retired. After the VMEXIT is handled properly by the hypervisor, the hypervisor will execute the VMRUN instruction. After all guest states are restored, the next instruction 1 in the guest VM will be executed and the instruction pointer will be advanced. Meanwhile, the attacker can set the APIC timer in the VMEXIT handler, so that the next APIC interrupt arrives when instruction 1 is executing (shown in blue lines in Fig. 3). Because of the second interrupt, another exception will be raised after instruction 1 is retired, which forces another VMEXIT and is trapped by attacker-controlled VMEXIT handler. In this way, the attacker can single-step the SEV VM.

In SEV-ES and SEV-SNP, the instruction pointer to be interrupted is encrypted and stored in an area called the VM Save Area (VMSA) during VMEXIT. Thus, the attacker cannot directly trace the execution of instructions. However, the attacker could still know whether the instruction pointer is advanced by monitoring the change of the encrypted instruction pointer inside VMSA. With the knowledge about the program context, the attacker knows the distance to the instruction in which the attacker is interested, therefore, allows the attacker to single-step the target instruction.

4.4 Power Interpolator

Limited by AMD's coarse-grained power consumption interface, a power interpolator is introduced to amplify the power consumption of the target instruction. PWRLEAK tests two interpolators: emulation-based and the existing interrupt-based interpolators. The emulation-based interpolator can be used to emulate a single instruction multiply times to amplify its power consumption, which is used by as an analysis tool to collect power consumption from a SEV VM. It's important to note that the emulation-based interpolator is only compatible with SEV VM and cannot be used with SEV-ES or SEV-SNP VM whose register states are encrypted during VMEXIT. The interrupt-based interpolator is a more general method that could be potentially applied to SEV-ES and SEV-SNP but with lots of noise. More discussion of their pros and cons, as well as their applicability in SEV-ES and SEV-SNP is covered in Sect. 6.

Emulation-Based Interpolator. Emulation-based interpolator modifies the kernel instruction emulation function to amplify the power consumption. KVM emulates the execution of the instruction that raises the exception in its handler using an emulation function to ensure that the same exception doesn't raise right after VMRUN. This function (x86_emulate_instruction) emulates the execution behavior of the instruction, for instance, accessing a specific memory location.

To deploy the emulation-based interpolator, PWRLEAK implements our own instruction emulation function by extending the x86_emulate_instruction function. In its original implementation, only instructions that access memory are emulated. We further extend the capability of the emulation function by emulating other instructions. Particularly, for those instructions that do not have memory access, we obtain the corresponding system states (*e.g.*, the value of registers) from the virtual machine control block (VMCB), and retrieve the instruction to be emulated using single-step and the program context. Finally, we execute the instruction directly in the hypervisor.

To apply this emulation-based interpolator for power analysis, PWRLEAK first hooks the VMEXIT handler. When the target instruction is interrupted, the VMEXIT trampoline handler is called. PWRLEAK forces the instruction to be emulated multiple times in the trampoline handler, surrounded by the instruction that reads the power consumption. The emulation-based interpolator can be used as a tool to analyze power-based vulnerabilities by precisely collecting the power consumption of specific instructions. Even the emulation-based approach works well in SEV, it cannot be recognized as an effective attack method because an attacker can directly read the values of registers in SEV, and such information leakage will cause more severe leaks. Meanwhile, this approach is limited by SEV-ES and SEV-SNP. SEV-ES and SEV-SNP would encrypt the Virtual Machine Save Area (VMSA), which stores all VM's state-related data. It would prevent the emulation-based interpolator from obtaining the register values. Therefore, the emulation-based interpolator cannot work with in the machine that supports SEV-ES and SEV-SNP.

Interrupt-Based Interpolator. The interrupt-based interpolator is considered to be a more general approach. In this paper, we have shown that it is feasible to use this approach in the baseline SEV, and similar approaches may also potentially work with SEV-ES or SEV-SNP. Specifically, by setting a value to the APIC timer, the attacker can control where the APIC interrupt arrives. The single-step makes the interrupt arrive when executing the first instruction after the VMRUN. Similar to the single-step, by setting a small APIC interval, the interrupt-based interpolator makes the interrupt arrive at the SEV VM within the VMRUN; thus, the exception will be raised before the next guest instruction has been executed, and the instruction pointer of the guest VM will not be advanced (shown in red lines in Fig. 3). In our experiment setup, the attacker conservatively underestimates the APIC interval and can directly verify whether the instruction is zero-stepped by checking the unencrypted rip register inside the SEV VM's VM control block. While this paper did not test for it, attackers may still be able to determine whether a zero-step or single-step occurred in SEV-ES or SEV-SNP environment through other side-channel information, such as observing changes in the ciphertext of the rip register, or by monitoring performance counters. With the interrupt-based interpolator approach, the attacker can force the VMRUN or target instruction in the VM to be executed multiple times for measurement purposes [30]. However, the root reason of distinguishable power consumption of an instruction amplified by interrupt-based interpolator is not verified by the paper. The power consumption difference could be introduced by different hamming weight in VMCB or be introduced by transient execution of the next instruction.

4.5 Power Attack

After amplifying and measuring the power consumption of the instructions with the power interpolator, PWRLEAK uses the power information to infer the secret. For operations with noticeable differences in power consumption, PWRLEAK uses the Simple Power Analysis (SPA) attack to infer the secret. PWRLEAK uses the cross correlation analysis (CCA) to infer the secret in those applications for which SPA fails.

5 Evaluation

In this section, we evaluate PWRLEAK using two case studies. We first present how to use only the emulation-based interpolator to analyze the power leakage from libjpeg, and then demonstrate the attack targeting an RSA implementation with the interrupt-based interpolator. The evaluation settings are identical as the experiment settings in Sect. 3, excepting the attacker now doesn't control the SEV-protected VM.

5.1 Infer Images from Libjpeg

```
1   int jpeg_idct_islow(j_decompress_ptr cinfo,
2   jpeg_component_info * compptr, JCOEFPTR coef_block,
3   JSAMPARRAY output_buf, JDIMENSION output_col) {
4       ...
5       /* Pass 1: process columns from input. */
6       inptr = coef_block;
7       for (ctr = DCTSIZE; ctr > 0; ctr--) {
8           if (inptr[DCTSIZE*1]==0 && inptr[DCTSIZE*2]==0 &&
9           inptr[DCTSIZE*3]==0 && inptr[DCTSIZE*4]==0 &&
10          inptr[DCTSIZE*5]==0 && inptr[DCTSIZE*6]==0 &&
11          inptr[DCTSIZE*7]==0) {
12          —— Simple Calculation ——
13          continue;}
14          —— Complex Calculation ——
15          }
16          /* Pass 2: process rows from work array. */
17          wsptr = workspace;
18          for (ctr = 0; ctr < DCTSIZE; ctr++) {
19   #ifndef NO_ZERO_ROW_TEST
20          if (wsptr[1]==0&&wsptr[2]==0&&wsptr[3]==0&&
21          wsptr[4]==0&&wsptr[5]==0&&wsptr[6]==0&&
22          wsptr[7]==0) {
23          —— Simple Calculation ——
24          continue; }
26   #endif
27          —— Complex Calculation ——
28          —— (with a lot of multiplication calculations) ——
29   }}
```

Fig. 4. IDCT function in libjpeg.

Libjpeg is a widely used image-rendering library that offers lossy image compression and decompression implementations. The input of the libjpeg library is a bitmap image. The decoding of a JPEG image transfer a bitmap image into blocks with 8x8 pixels with three steps: *decompression, dequantization*, and *inverse discrete cosine transformation* (IDCT). The encoding procedure transfers blocks to a bitmap image with *discrete cosine transform, quantization*, and *compression*. The JPEG image is shown on the screen based on the decoded pixels. With enough information about each 8x8 pixels, adversaries can recover the whole JPEG image.

In IDCT algorithm [28], there are two loops to handle a block (*i.e.* eight columns and eight rows). A simple calculation applies when all elements in a row or a column are zeros; otherwise, a complex calculation with more page faults applies. The attacker can then infer the value of each block by normalizing the number of data-page faults. To mitigate this vulnerability, libjpeg (version 6b, Fig. 4) implemented the flag NO_ZERO_ROW_TEST. When the flag is enabled, all rows use complex calculation, thus page faults can not infer data in rows. Thus increases the difficulty of using only page fault information to recover JPEG images.

In this experiment, we demonstrate that the power information can further be exploited to infer rows in JPEG images on the IDCT implementation of the newest libjpeg library. We first present that the power-based attack is also useful when the program is vulnerable to order-based attacks. Using the emulation-based interpolator to measure the power consumption of the simple and the complex calculations (wherein the input is a column with four pixels equals 0) and using the SPA to analyze the results. Particularly, we amplified some target instructions 100, 000 times, measured the power consumption, and inferred the instruction and the secret. The results are presented in Fig. 5(a), which indicates that the power of the simple and complex calculations can be easily distinguished with the emulation-based interpolator.

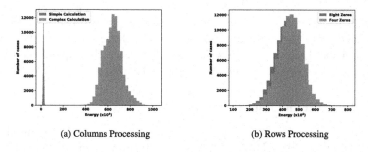

(a) Columns Processing   (b) Rows Processing

Fig. 5. Energy Consumption of Emulation-based Interpolator.

Then, we demonstrate the power information is another covert channel when order-based attacks is mitigated. To analyze the energy consumption of rows of JPEG images, we use emulation-based approach to simulate/collect the power consumption. We measured the power consumption of each instruction 100, 000 times and amplified each one 100, 000 times with the emulation-based interpolator. Finally, we applied the SPA to analyze the results. The results are presented in Fig. 5(b), where the blue bars are the rows with all bits (except the first one) as zeros, and the orange bars are the rows with the Hamming weight equal to 4. The result indicates that these two conditions are discernible. This is because one of the most energy-consuming calculations is multiplication, which consumes much less energy when multiplying by 0. While this paper did not reconstruct the JPEG image, attackers may still be able to recover the whole image as the columns and rows with all bits as zeros is discernible [50].

5.2   Steal Private Exponent in RSA

RSA is a widely used asymmetric cryptographic algorithm. Modular exponentiation is one of the most important components of the RSA algorithm. To mitigate attacks such as SPA [32] and DPA [22], the modular exponentiation algorithm with message blinding [14,43] was discovered (as shown in Algorithm 1) by introducing a random variable. However, attackers could still exploit this algorithm when they can discover the correlation between the private exponent and the power consumption.

Here we targeted at a non-constant time RSA implementation of Intel's Integrated Performance Primitives (Intel IPP) library [19] with modular exponentiation algorithm with message blinding, and we exploited both emulation-based and interrupt-based interpolators to infer the private exponent. The RSA implementation in the IPP library first calls the `ippsRSA_Decrypt/ippsRSA_Encrypt` function and then selects the actual function (e.g., `gsRSAprv_cipher`) for the encryption and the decryption based on the instruction set supported by the CPU. We evaluated the effect of the power-based attacks on it with a 512-bit RSA private exponent.

Interrupt-based Interpolator. Firstly, we try to exploit the interrupt-based interpolator on a modular exponentiation algorithm with message blinding. As the algorithm runs inside the SEV, neither plaintext nor ciphertext are available to adversaries. Thus, the CCA attack [5,33] is selected. The underlying assumption of the CCA attack is that the correlation coefficient of the power consumption between two "$z = z * T[d_i]$ mod n" operations would be higher if the values of two $T[d_i]$s are the same. Otherwise, the correlation coefficient would be relatively low.

Algorithm 1: Modular Exponentiation

Input: x, n, d=$(d_{e-1}, d_{e-2}, \cdots, d_2, d_1, d_0)$,
a=$r - 1$ mod n (r is a random number)
Output: x^d mod n
begin
    T[0] $\leftarrow$ a, T[1] $\leftarrow$ $x*r$ mod n, $z \leftarrow r$ mod x
    foreach $i = e - 1$ *to 0* **do**
        | $z \leftarrow z*z$ mod n, $z \leftarrow z*T[d_i]$ mod n
    end
    z=z*a mod n
    return z
end

Power Trace Collection. For verification purposes, we use a relatively short RSA private exponent (512-bit) to decrypt a ciphertext in this experiment. To collect the power traces of the RSA operation, we focus on the "$z = z * T[d_i]$ mod n" operation in each iteration. For each instruction of this operation, we apply the interrupted-based interpolator to amplify the power consumption of each instruction N times (zero-stepping). As N increases, the precision of instruction's power consumption also becomes greater. Then, we organize all the power information collected into our defined format (as shown in Definition 1). In total, $3,000,000$ power traces were collected.

Definition 1. *Let $\mathcal{P}i = \{\mathcal{P}i_0, \mathcal{P}i_1, \cdots, \mathcal{P}i_r\}$ be the power trace of i_{th} execution of the RSA decryption operation, and in total r power traces collected. $\mathcal{P}i_j = \{e_{i,j,0}, e_{i,j,1}, \cdots, e_{i,j,k}\}$ corresponding to the power information of the j_{th} iteration of the multiplication operation ("$z = z * T[d_i]$ mod n") for the i_{th} power trace. This multiplication operation has k instructions, and $e_{i,j,k}$ indicates the power consumption of the k_{th} instruction in it.*

Correlation Calculation. To calculate the correlation coefficient, we first randomly select a jth bit as the reference bit. Then we calculate the Pearson correlation coefficient of power consumption between the jth bit and all other bits with the equation shown in Eq. 1 [49]. The result of this equation is the correlation coefficient of the power consumption for the same instruction between different bits in the private exponent.

$$\rho(\mathcal{P}i_{j1},\mathcal{P}i_{j2}) = \frac{\sum_{i=0}^{r-1}(e_{i,j1,1}e_{i,j2,1}) - \frac{\sum_{i=0}^{r-1}e_{i,j1,1}\sum_{i=0}^{r-1}e_{i,j2,1}}{r}}{\sqrt{(\sum_{i=0}^{r-1}e_{i,j1,1}^2 - \frac{(\sum_{i=0}^{r-1}e_{i,j1,1})^2}{r})}\sqrt{(\sum_{i=0}^{r-1}e_{i,j2,1}^2 - \frac{(\sum_{i=0}^{r-1}e_{i,j2,1})^2}{r})}}$$

(1)

An example of this algorithm is shown in Fig. 6. Each column is the power consumption of an instruction in the operation "$z = z * T[d_i] \mod n$". For instance, the first red column is the power consumption of the first instruction in the operation when processing bit 0, and the second red column is the power consumption of the first instruction when processing bit j. We calculate the power correlation between the same instruction in different bits; thus, 512 correlations are calculated for each instruction. When the operation "$z = z*T[d_i] \mod n$" consists of k instructions, $512*k$ correlations are calculated in total.

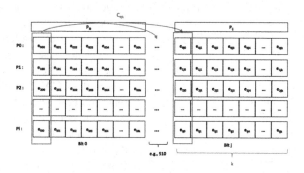

Fig. 6. Correlation Calculation Algorithm.

Exploit Key Bits. We use the correlation coefficient to distinguish bits in the private exponent [46]. In particular, a higher correlation coefficient indicates that two bits (*e.g.*, $T[0]$ or $T[1]$) are the same. A relatively low correlation coefficient means that either two bits are different, or a significant amount of noise is included in the power information. As the noise introduced by the APIC interrupt and the system (*e.g.*, random time delay, random clock, *etc.*) could affect the result of the correlation coefficient, we only keep those instructions that have a high correlation coefficient. In particular, the following two steps are used:

- We filter the noise with an intermediate value, the variance. As shown in Fig. 7a, each row consists of correlation values of the same instruction from various bits. We first calculate the variance for each instruction (each row), then keep those instructions with a relatively higher variance. A higher variance (peaks in Fig. 7b) means that the power information of this instruction can help infer the private exponent.
- Then, we add up the correlation coefficient of these selected instructions (*e.g.*, rows in red in Fig. 7a) for each bit, then deduce the private exponent based on the value of the summed correlation coefficients with the threshold-based approach. A higher sum of the correlation coefficient means that this bit has a higher chance to be the

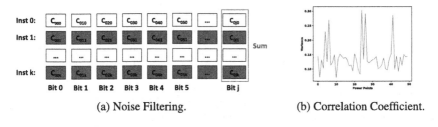

(a) Noise Filtering.   (b) Correlation Coefficient.

Fig. 7. Interrupt-based Interpolator.

same as the reference bit. A lower correlation indicates that this bit has higher probability to be opposite to the reference bit. Thus, an attacker can use correlation analysis to infer the value of each bit and have different levels of confidence in the predicted value of each bit.

Evaluation Results. We randomly generated a 512-bit RSA key using openssl and used the two interpolators discussed above to recover the private exponent. To recover this RSA key (private exponent), we recorded $3,000,000$ traces in total.

Interrupted-Based Interpolator. The interrupted-based interpolator method could correctly infer 427 out of 512 bits of the private exponent by comparing with the correct key. With only correlation values, it is hard to know which bits of the key do not recover correctly. When comparing this result with other works with a recovery rate greater than 90% [5], the lower recovery rate of the interrupt-based interpolator indicates that this approach is affected by some noise. Such noise might be averaged out with the increasing number of traces collected, we leave this for future work. However, the time needed for collecting power consumption data also increases with the number of traces, and in the experiment described in the paper, collecting 3 million traces took around 80 h.

Emulation-Based Interpolator. With the same approach discussed above, we measured the power consumption of each instruction in the operation "$z \leftarrow z*\text{T}[d_i] \bmod n$" with the emulation-based interpolator. In particular, for the evaluation purpose, we generated the RSA key pair with a 512-bit modulus. For each instruction, we applied the emulation-based interpolator to amplify the power consumption $100,000$ times. In total, $1,000,000$ different power traces were collected. After applying the CCA attack, we successfully recovered the private exponent without an error.

6   Discussion

Countermeasures. The power-based side-channel attack needs to gather fine-grained power information during run-time in order to analyze and infer secret using the collected data. Thus, adding additional noise may be one potential way to prevent power-based side-channel attacks. The power-based side-channel attack could be prevented if the hardware or the VM itself could add noises during run-time to hide the power consumption pattern caused by different instructions and operands. The hardware manufacturers can also use a microcode patch to disable the hardware reporting interfaces of TEE's power assumption to prevent such attacks. For example, affected by two general power-based side-channel attacks [29,30], AMD has added restrictions on accessing power-consumption interfaces in newer kernel versions, and Linux has also removed some related drivers that could potentially cause leakage [34].

Comparison with Other Power Attacks. To launch a power-based side-channel attack, the CPA attack is a widely used method [9,10,31], which can resist noise. However, CPA attack mainly targets at algorithms and requires either plaintext or ciphertext, which is not the case in the AMD SEV's scenario. The most related work to us is Platypus [30], which uses Intel RAPL to break Intel SGX protection, steal private keys from constant-time cryptographic implementation (AES-NI), and study power consumption of instructions in AMD platform. Considering that different CPU hardware and TEE design (Intel SGX aims at protecting an application instead of a VM) could introduce different power pattern, PwrLeak could be a complementary work to Platypus with similar approaches but target a different TEE design and cryptographic implementation with low-secure level (non-constant time Intel IPP library).

Future Work. Due to equipment limitations, we did not perform experiments on SEV-ES and SEV-SNP. Here we discuss the feasibility of power side-channel attacks on these machines and treat them as future work. The current version of the emulation-based interpolator could not work on SEV-ES and SEV-SNP due to the encrypted VMSA. The interrupt-based interpolator could potentially work on both SEV-ES and SEV-SNP. However, this approach would encounter a substantial amount of unstable noise introduced by additional protection from SEV-ES or SEV-SNP. For example, in SEV-ES, there is an integrity check for the VM Save Area region (the region used to encrypt and backup registers) during each VMEXIT or VMRUN. In SEV-SNP, each memory write access in the case of a TLB miss introduces a Reverse Map Table (RMP) check. These additional protections may introduce inaccuracies in the observed energy consumption. Therefore, a precise noise cancellation algorithm or an amplifier that can further magnify the difference in power consumption may be necessary for such side-channel attacks to work in SEV-ES or SEV-SNP VMs.

7   Related Work

Other Attacks Against AMD SEV. AMD SEV has been studied by both industry and academia since its first release in 2016. Faced with a strong threat model in which the entire software stack is not trusted, previous work showed that AMD SEV suffers from numerous attack surfaces, including both *incomplete system designs* [15,18,24,27,38,48] and *side-channel attacks* [23,26,47]. For *incomplete system designs*, AMD actively addresses existing attacks by providing microcode patches [7] and adding new hardware extensions (including AMD SEV Encrypted States (SEV-ES) [21] and AMD SEV Secure Nested Paging (SEV-SNP) [6]) in addition to the baseline AMD SEV. However, for *side-channel attacks*, which are not included in AMD SEV's thread model and indirectly leak secret from the SEV-protected VM [23,25,26], AMD typically does not provide fixes for such attacks. Common side-channel attacks in AMD SEV include page table-based side-channel attacks [39], cache side-channel attacks, PMC-based side-channel attacks [47], and ciphertext side-channel attacks [23]. The defense mechanisms against such side-channel attacks often involve refactoring source code to avoid certain patterns or gadgets, or adopting code with constant-time implementations.

8 Conclusion

In this paper, we have demonstrated the potency of power-based side-channel attacks in extracting secrets from AMD SEV-protected VMs. Through a series of exploratory experiments and an emulation-based interpolator, we show that adversaries can still notice the differences in the instruction and operand level with the 1 ms coarse-grained power sampling interval provided by AMD. Additionally, we have successfully leaked a random generated RSA key in an IPP implementation using PWRLEAK.

Acknowledgments. We would like to thank the anonymous reviewers and the shepherd Moritz Lipp for their very helpful comments and feedback during revision, which have significantly improved the quality and clarity of the work. This research was partially supported by NSF award 2207202. Any opinions, findings, and conclusions or recommendations in this paper are those of the authors and do not necessarily reflect the views of the NSF.

References

1. Confidential computing: an AWS perspective (2021). https://aws.amazon.com/blogs/security/confidential-computing-an-aws-perspective/, 2021. Aug, 2021
2. SEV Secure Nested Paging Firmware ABI Specification (2021). https://www.amd.com/system/files/TechDocs/56860.pdf
3. Arm Confidential Compute Architecture (2022). https://www.arm.com/architecture/security-features/arm-confidential-compute-architecture. 2022. Dec, 2022
4. Intel trust domain extensions (2022). https://www.intel.com/content/www/us/en/developer/articles/technical/intel-trust-domain-extensions.html. 2022. Dec, 2022
5. Kuzu, E.A., Soysal, B., Şahinoğlu, M., Güvenç, U., Tangel, A.: New cross correlation attack methods on the montgomery ladder implementation of rsa. In: 2013 3rd IEEE International Advance Computing Conference (IACC), pp. 138–142 (2013)
6. AMD. AMD SEV-SNP: Strengthening VM isolation with integrity protection and more. White paper (2020)
7. AMD. AMD Secure Encryption Virtualization (SEV) Information Disclosure (Bulletin ID: AMD-SB-1013) (2021). https://www.amd.com/en/corporate/product-security/bulletin/amd-sb-1013
8. AMD. AMDSEV branch (2022). https://github.com/AMDESE/AMDSEV/
9. Bottinelli, P., Bos, J.W.: Computational aspects of correlation power analysis. J. Cryptographic Eng. **7**(3), 167–181 (2017)
10. Brier, E., Clavier, C., Olivier, F.: Correlation power analysis with a leakage model. In: Joye, M., Quisquater, J.-J. (eds.) CHES 2004. LNCS, vol. 3156, pp. 16–29. Springer, Heidelberg (2004). https://doi.org/10.1007/978-3-540-28632-5_2
11. Chari, S., Rao, J.R., Rohatgi, P.: Template attacks. In: Kaliski, B.S., Koç, K., Paar, C. (eds.) CHES 2002. LNCS, vol. 2523, pp. 13–28. Springer, Heidelberg (2003). https://doi.org/10.1007/3-540-36400-5_3

12. Chen, Y., Jin, X., Sun, J., Zhang, R., Zhang, Y.: Powerful: mobile app fingerprinting via power analysis. In: IEEE INFOCOM 2017-IEEE Conference on Computer Communications, pp. 1–9. IEEE (2017)
13. Clavier, C., Feix, B., Gagnerot, G., Roussellet, M., Verneuil, V.: Improved collision-correlation power analysis on first order protected AES. In: Preneel, B., Takagi, T. (eds.) CHES 2011. LNCS, vol. 6917, pp. 49–62. Springer, Heidelberg (2011). https://doi.org/10.1007/978-3-642-23951-9_4
14. Coron, J.-S.: Resistance against differential power analysis for elliptic curve cryptosystems. In: Koç, Ç.K., Paar, C. (eds.) CHES 1999. LNCS, vol. 1717, pp. 292–302. Springer, Heidelberg (1999). https://doi.org/10.1007/3-540-48059-5_25
15. Du, Z.-H., et al.: Secure encrypted virtualization is unsecure. arXiv preprint arXiv:1712.05090 (2017)
16. Eisenbarth, T., Paar, C., Weghenkel, B.: Building a side channel based disassembler. In: Gavrilova, M.L., Tan, C.J.K., Moreno, E.D. (eds.) Transactions on Computational Science X. LNCS, vol. 6340, pp. 78–99. Springer, Heidelberg (2010). https://doi.org/10.1007/978-3-642-17499-5_4
17. Google. Introducing google cloud confidential computing with confidential VMs (2020). https://cloud.google.com/blog/products/identity-security/introducing-google-cloud-confidential-computing-with-confidential-vms
18. Hetzelt, F., Buhren, R.: Security analysis of encrypted virtual machines. ACM SIGPLAN Notices 52(7), 129–142 (2017)
19. Intel integrated performance primitives. https://software.intel.com/content/www/us/en/develop/tools/oneapi/components/ipp.html
20. Itoh, K., Yamamoto, D., Yajima, J., Ogata, W.: Collision-based power attack for RSA with small public exponent. IEICE Trans. Inf. Syst. 92(5), 897–908 (2009)
21. Kaplan, D.: Protecting VM register state with SEV-ES. White paper (2017)
22. Kocher, P., Jaffe, J., Jun, B., et al.: Introduction to differential power analysis and related attacks (1998)
23. Li, M., Wilke, L., Wichelmann, J., Eisenbarth, T., Teodorescu, R., Zhang, Y.: A systematic look at ciphertext side channels on AMD SEV-SNP. In: 2022 IEEE Symposium on Security and Privacy (SP), pp. 1541–1541. IEEE Computer Society (2022)
24. Li, M., Zhang, Y., Lin, Z.: Crossline: Breaking "security-by-crash" based memory isolation in AMD SEV. In: Proceedings of the 2021 ACM SIGSAC Conference on Computer and Communications Security, pp. 2937–2950 (2021)
25. Li, M., Zhang, Y., Lin, Z., Solihin, Y.: Exploiting unprotected I/O operations in AMD's secure encrypted virtualization. In: 28th USENIX Security Symposium (USENIX Security 19), pp. 1257–1272 (2019)
26. Li, M., Zhang, Y., Wang, H., Li, K., Cheng, Y.: CIPHERLEAKS: breaking constant-time cryptography on AMD SEV via the ciphertext side channel. In: 30th USENIX Security Symposium (USENIX Security 21), pp. 717–732 (2021)
27. Li, M., Zhang, Y., Wang, H., Li, K., Cheng, Y.: TLB Poisoning Attacks on AMD Secure Encrypted Virtualization. In: Annual Computer Security Applications Conference, pp. 609–619 (2021)
28. Libjpeg. Libjpeg version 6b Files. https://sourceforge.net/projects/libjpeg/files/libjpeg/6b/
29. Lipp, M., Gruss, D., Schwarz, M.: AMD prefetch attacks through power and time. In: 31st USENIX Security Symposium (USENIX Security 22), pp. 643–660 (2022)
30. Lipp, M., et al.: Platypus: software-based power side-channel attacks on x86. In: 2021 IEEE Symposium on Security and Privacy (SP), pp. 355–371. IEEE (2021)
31. Lo, O., Buchanan, W.J., Carson, D.: Power analysis attacks on the AES-128 S-box using differential power analysis (DPA) and correlation power analysis (CPA). J. Cyber Secur. Technol. 1(2), 88–107 (2017)

32. Mangard, S., Oswald, E., Popp, T.: Power analysis attacks: revealing the secrets of smart cards, vol. 31. Springer Science & Business Media (2008)
33. Messerges, T.S., Dabbish, E.A., Sloan, R.H.: Power analysis attacks of modular exponentiation in smartcards. In: Koç, Ç.K., Paar, C. (eds.) CHES 1999. LNCS, vol. 1717, pp. 144–157. Springer, Heidelberg (1999). https://doi.org/10.1007/3-540-48059-5_14
34. Larabel, M.: AMD Energy Driver Booted From The Linux 5.13 Kernel. https://www.phoronix.com/news/Linux-5.13-AMD-Energy-Removed (2021)
35. Michalevsky, Y., Schulman, A., Veerapandian, G.A., Boneh, D., Nakibly, G.: PowerSpy: location tracking using mobile device power analysis. In: 24th USENIX Security Symposium (USENIX Security 15), pp. 785–800 (2015)
36. Microsoft. Azure and AMD announce landmark in confidential computing evolution (2021). https://azure.microsoft.com/en-us/blog/azure-and-amd-enable-lift-and-shift-confidential-computing/
37. Moradi, A., Mischke, O., Eisenbarth, T.: Correlation-enhanced power analysis collision attack. In: Mangard, S., Standaert, F.-X. (eds.) CHES 2010. LNCS, vol. 6225, pp. 125–139. Springer, Heidelberg (2010). https://doi.org/10.1007/978-3-642-15031-9_9
38. Morbitzer, M., Huber, M., Horsch, J.: Extracting secrets from encrypted virtual machines. In: Proceedings of the Ninth ACM Conference on Data and Application Security and Privacy, pp. 221–230 (2019)
39. Morbitzer, M., Huber, M., Horsch, J., Wessel, S.: Severed: subverting AMD's virtual machine encryption. In: Proceedings of the 11th European Workshop on Systems Security, pp. 1–6 (2018)
40. Niu, Y., Zhang, J., Wang, A., Chen, C.: An efficient collision power attack on AES encryption in edge computing. IEEE Access 7, 18734–18748 (2019)
41. Park, J., Xu, X., Jin, Y., Forte, D., Tehranipoor, M.: Power-based side-channel instruction-level disassembler. In: 2018 55th ACM/ESDA/IEEE Design Automation Conference (DAC), pp. 1–6. IEEE (2018)
42. Strobel, D., Bache, F., Oswald, D., Schellenberg, F., Paar, C.: Scandalee: a side-channel-based disassembler using local electromagnetic emanations. In: 2015 Design, Automation & Test in Europe Conference & Exhibition (DATE), pp. 139–144. IEEE (2015)
43. Sung-Ming, Y., Kim, S., Lim, S., Moon, S.: A countermeasure against one physical cryptanalysis may benefit another attack. In: Kim, K. (ed.) ICISC 2001. LNCS, vol. 2288, pp. 414–427. Springer, Heidelberg (2002). https://doi.org/10.1007/3-540-45861-1_31
44. Van Bulck, J., Piessens, F., Strackx, R.: SGX-step: a practical attack framework for precise enclave execution control. In: Proceedings of the 2Nd Workshop on System Software for Trusted Execution, (SysTEX'17) (2017)
45. Vermoen, D., Witteman, M., Gaydadjiev, G.N.: Reverse engineering Java Card applets using power analysis. In: Sauveron, D., Markantonakis, K., Bilas, A., Quisquater, J.-J. (eds.) WISTP 2007. LNCS, vol. 4462, pp. 138–149. Springer, Heidelberg (2007). https://doi.org/10.1007/978-3-540-72354-7_12
46. Wan, W., Yang, W., Chen, J.: An optimized cross correlation power attack of message blinding exponentiation algorithms. China Commun. 12(6), 22–32 (2015)
47. Werner, J., Mason, J., Antonakakis, M., Polychronakis, M., Monrose, F.: The SEVerESt of them all: inference attacks against secure virtual enclaves. In: Proceedings of the 2019 ACM Asia Conference on Computer and Communications Security, pp. 73–85 (2019)
48. Wilke, L., Wichelmann, J., Morbitzer, M., Eisenbarth, T.: Sevurity: No security without integrity: breaking integrity-free memory encryption with minimal assumptions. In: 2020 IEEE Symposium on Security and Privacy (SP), pp. 1483–1496. IEEE (2020)
49. Witteman, M.F., van Woudenberg, J.G.J., Menarini, F.: Defeating RSA multiply-always and message blinding countermeasures. In: Kiayias, A. (ed.) CT-RSA 2011. LNCS, vol. 6558, pp. 77–88. Springer, Heidelberg (2011). https://doi.org/10.1007/978-3-642-19074-2_6

50. Xu, Y., Cui, W., Peinado, M.: Controlled-channel attacks: Deterministic side channels for untrusted operating systems. In: Proceedings of the 2015 IEEE Symposium on Security and Privacy (SP'15). IEEE (2015)
51. Yang, S., Wolf, W., Vijaykrishnan, N., Serpanos, D.N., Xie, Y.: Power attack resistant cryptosystem design: A dynamic voltage and frequency switching approach. In: Design, Automation and Test in Europe, pp. 64–69. IEEE (2005)
52. Zeichick, A.: Security Ahoy! Flying the NX Flag on Windows and AMD64 To Stop Attacks. Advanced Micro Devices, March 2007
53. Zhao, B., Wang, L., Jiang, K., Liang, X., Shan, W., Liu, J.: An improved power attack on small RSA public exponent. In: 2016 12th International Conference on Computational Intelligence and Security (CIS), pp. 578–581. IEEE (2016)

Security and Machine Learning

MADVEX: Instrumentation-Based Adversarial Attacks on Machine Learning Malware Detection

Nils Loose[(✉)], Felix Mächtle, Claudius Pott, Volodymyr Bezsmertnyi,
and Thomas Eisenbarth

Institute for IT-Security, Ratzeburger Allee 160, 23562 Lübeck, Germany
{n.loose,f.maechtle,c.pott,thomas.eisenbarth}@uni-luebeck.de,
volodymyr.bezsmertnyi@student.uni-luebeck.de

Abstract. WebAssembly (`Wasm`) is a low-level binary format for web applications, which has found widespread adoption due to its improved performance and compatibility with existing software. However, the popularity of `Wasm` has also led to its exploitation for malicious purposes, such as cryptojacking, where malicious actors use a victim's computing resources to mine cryptocurrencies without their consent. To counteract this threat, machine learning-based detection methods aiming to identify cryptojacking activities within `Wasm` code have emerged. It is well-known that neural networks are susceptible to adversarial attacks, where inputs to a classifier are perturbed with minimal changes that result in a crass misclassification. While applying changes in image classification is easy, manipulating binaries in an automated fashion to evade malware classification without changing functionality is non-trivial. In this work, we propose a new approach to include adversarial examples in the code section of binaries via instrumentation. The introduced gadgets allow for the inclusion of arbitrary bytes, enabling efficient adversarial attacks that reliably bypass state-of-the-art machine learning classifiers such as the CNN-based MINOS recently proposed at NDSS 2021. We analyze the cost and reliability of instrumentation-based adversarial example generation and show that the approach works reliably at minimal size and performance overheads.

Keywords: Malware Detection · Adversarial Attack · Binary Instrumentation · MINOS · Cryptojacking

1 Introduction

With the introduction of WebAssembly (`Wasm`) in 2017, web applications are able to utilize a system's CPUs with near-native efficiency [1]. `Wasm` allows developers to make computationally heavy applications available in-browser and has since been used for games, text processing, visualizations, and media players [14,21]. On the downside, malicious parties have also utilized `Wasm` to distribute malicious binaries to victims that visit an infected website and thus gain access to the victim's resources without having to gain access to their system. In particular, the near-native performance of `Wasm` and the support provided by all

D. Gruss et al. (Eds.): DIMVA 2023, LNCS 13959, pp. 69–88, 2023.
https://doi.org/10.1007/978-3-031-35504-2_4

major browsers make WebAssembly a prime target for cryptojacking attacks [14,21,35]. In-browser cryptojacking or drive-by cryptocurrency mining allows an attacker to utilize their victim's computational resources for mining cryptocurrencies without their knowledge or consent, thus profiting from the returns without having to pay for the spent energy. To address this issue, various methods have been proposed to protect against cryptojacking attacks. However, while fast, traditional static approaches like blacklisting malicious hosts or matching signatures are easily bypassed [31]. Dynamic detection systems [15,17,30], on the other hand, rely on more sophisticated metrics that cause a runtime overhead and require the malicious binary to be executed. MINOS, a lightweight machine learning-based detection system, provides a promising solution to this problem [23]. By transforming Wasm binaries to grey-scale images, MINOS can utilize a convolutional neural network (CNN) for the classification of binaries. This provides a rapid and effective approach that can be applied prior to executing the binaries, thereby offering efficient protection against in-browser cryptojacking attacks. While promising, CNNs are known to be susceptible to adversarial attacks [39]. Malicious parties looking to distribute their malware have a high incentive to evaluate possible avenues for bypassing detection frameworks. In particular, the development of more sophisticated evasion techniques by attackers could render existing detection methods ineffective. Adversarial examples are usually crafted under the assumption that small changes to the input are neglectable. However, applying adversarial examples to binaries that follow strict syntactical and semantical rules requires specific placement of adversarial payloads without invalidating the binary or changing the semantics. Still, attacks leveraging adversarial examples to bypass visualization-based malware detectors have been proven to succeed on Windows Portable Executables [16,20,28].

In this paper, we evaluate the feasibility of utilizing adversarial examples against the Wasm-based classifier MINOS [23] presented at NDSS 2021. We demonstrate the feasibility of inserting semantic-preserving gadgets using binary instrumentation into the code section of WebAssembly applications, allowing effective crafting of adversarial examples inside the gadget, thus enabling the evasion of the MINOS detection system. In contrast to existing work, we add the adversarial payload directly into the application's control flow and introduce both size-efficient (SE) and optimization-resistant (OR) gadgets. Our findings shed light on the potential weaknesses of machine learning-based classifiers in detecting cryptojacking and highlight the need for ongoing efforts to improve their robustness and security, particularly when classifiers are applied in scenarios with incentives to evade classification. To summarize, our key contributions are:

- Comprehensive collection of malign Wasm samples from the *Cisco Umbrella 1 Million* websites list.
- A novel approach for automatically crafting adversarial examples in code by introducing semantic-preserving instruction gadgets via instrumentation.
- Demonstrating a grey-box adversarial attack against the MINOS classifier by training a substitute model and applying our gadgets.
- A comprehensive evaluation of the efficacy and costs of the attack.

2 Background

2.1 WebAssembly

WebAssembly (Wasm) [1] is a binary instruction format for a stack-based virtual machine that enables high-performance applications that run seamlessly in web browsers. It is designed to provide near-native performance to web applications and allows developers to write applications in various programming languages, including C, C++, and Rust, while still being executed in the browser. Wasm is supported by all major web browsers and has gained significant traction in recent years, particularly in resource-intensive applications, where the performance benefits provided by Wasm are especially important. In most settings, Wasm is integrated into the JavaScript code of a website, from where the Wasm modules are loaded, and the respective functions are called. Its stack-based architecture, widespread support, and versatility make it an essential tool for modern web development.

2.2 Cryptojacking Malware

Cryptocurrency mining is the process of solving complex mathematical problems in order to validate transactions and add new blocks to a blockchain network [22]. The process requires a significant amount of computational power and energy. As compensation for the computation time, miners are rewarded with new units of the respective cryptocurrency. This reward mechanism is a key component of the decentralized nature of many cryptocurrencies, as it incentivizes individuals and organizations to participate in the network and maintain its security. However, as the difficulty of mining increases and the competition among miners grows, the margin between the resources spent on mining and the returned profits diminishes. If a malicious actor manages to utilize a victim's resources for mining, the computational cost is removed from the equation. In general, the unauthorized use of a device's computing power to mine cryptocurrencies, typically without the knowledge or consent of the device's owner, is referred to as *cryptojacking*. This type of attack can occur via host- or browser-based mining and can have significant impacts on both individual users and organizations. Host-based cryptojacking requires the installation of a cryptocurrency miner on the victim's machine through, i.e., malicious software installed by the victim [35]. Browser-based cryptojacking is a method of exploiting a victim's device through a malicious website. The attacker inserts a script into the website's code that runs in the victims' browser upon visiting the site and uses their device's processing power to mine cryptocurrency while profiting the owner of the operation. With the introduction of WebAssembly and its near-native speed, the efficiency of browser-based mining has significantly increased, making the attack lucrative. Unlike traditional malware, browser-based cryptojacking does not require the victims to download any files, making it subtle and difficult to prevent.

2.3   Malware Detection

Identifying whether a binary contains malicious functionality is an active area of research across different types of binaries. Various approaches have been proposed for detecting *cryptojacking*, one of the primary malicious usages of Wasm binaries [21]. Due to the reliance of cryptojacking malware on network communication, network-based detection systems have been proposed, analysing the network traffic [32]. Host-based detection frameworks rely, in general, on either static or dynamic analysis to identify malware. Dynamic approaches observe the execution of a binary while monitoring key metrics such as memory consumption [25], the number of executed arithmetic operations [37], or through CPU profiling [17]. Prevention techniques that identify malware based on resource consumption can be circumvented through throttling [12]. Additionally, a number of machine learning classifiers have been proposed that require dynamic features such as API calls and resource information [30] or runtime information such as the number of web sockets or workers [15]. In order to generate dynamic features, the potentially malicious binary needs to be executed on the host's machine. Static approaches, on the other hand, do not require the evaluated code to be executed; instead, the binary is directly evaluated, for example, by matching known signatures or URL blacklisting [12]. However, these techniques can be circumvented using obfuscation [31]. MincrRay [31] relies on the static detection of hash semantics to make obfuscation-based prevention harder as the semantics of the functions are evaluated.

In general, efficiently detecting whether a WebAssembly binary utilizes the host's resources for mining cryptocurrencies without relying on dynamic features allows a detection framework to warn the user that a malicious binary is loaded before the execution of the binary. Nassem *et al.* developed MINOS [23], a lightweight real-time detection system that aims to efficiently detect whether a WebAssembly binary utilizes the host's resources for cryptomining using a CNN. MINOS is designed to be implemented as a browser plug-in which uses the detection framework to warn users about any detected cryptomining binaries before they are executed. Upon visiting a website that loads a Wasm binary, the detection framework transforms the bytes contained inside the binary into a two-dimensional grey-scale image which is then evaluated by a pre-trained CNN. This architecture allows the system to classify a binary, on average, in 25.9 *ms* while achieving an overall accuracy of 98.97% against an in-the-wild dataset [23].

2.4   Adversarial Attacks

Deep neural networks, along with other machine learning models, have been discovered to be susceptible to adversarial attacks on their input data [4,34]. Given a target model θ, an input x and a target class $t \neq \theta(x)$, an adversaries objective is to find a minimal perturbation δ_x under a norm $\mathcal{N} = || \cdot ||$ s.t.

$$\theta(x + \delta_x) = t \tag{1}$$

Minimizing the perturbation vector δ_x under a norm $\mathcal{N}$ ensures that the original input x and the newly generated input, or *adversarial example*, $x^* = x + \delta_x$ are close to each other under a given distance metric $\mathcal{D}$. However, finding a perturbation δ_x that satisfies Eq. 1 is generally a hard problem due to the nonlinearity of the evaluated model θ [34]. Existing methods for crafting an adversarial example, such as the L-BFGS, solve the problem using approximations [39]. Carlini and Wagner (C&W) proposed a different approach by transforming the constraint shown in Eq. 1 into an optimization problem using an appropriately chosen objective function $\mathcal{L}$, s.t. if $\theta(x + \delta_x) = t$ is satisfied, $\mathcal{L}(x^*) \leq 0$ holds [8]. By moving the constraint into the minimization term, the problem of finding an adversarial example is an optimization task that minimizes $\mathcal{N}(\delta_x) + \epsilon \cdot \mathcal{L}(x^*)$ such that $x^* \in [0,1]^n$ where $\epsilon > 0$ is a suitably chosen constant. The optimization problem is solved using gradient-based optimization methods [5]. The gradient of the objective function with respect to the input x is used to update the perturbation δ_x in each iteration of the optimization process. The process is repeated until the minimum perturbation, which results in the adversarial example being classified as the target class t, is found. Without access to the gradients of the target model θ, the aforementioned attack cannot be utilized. However, given query access, the adversary can train a local substitute network [27] by querying the target classifier with synthesized or otherwise gathered data. Using the results obtained through inference against the target network as labels, the local model is trained. Due to the *transferability* between models, it is possible to train a machine learning model that mimics the behaviour of a target model [13]. In a black-box scenario [27], a network with unknown architecture is attacked, requiring a custom architecture for the local substitute network. In the grey-box scenario, additional information about the target network, such as parameters or its architecture, is known, and hence the substitution network architecture can be chosen similarly to the target model. The local model can then be utilized to generate adversarial examples that are transferable to the target network [27].

3   MADVEX: Crafting Functional Adversarial Binaries

The MINOS classifier [23] uses an image-based machine learning technique to quickly identify malicious WebAssembly binaries. However, such classifiers are shown to be vulnerable to adversarial attacks [34]. This section describes the attack methodology used to craft binaries that are misclassified by MINOS. To illustrate the applicability of such an attack, we limit the adversary and assume a grey-box scenario where the attacker has query access to the model and knowledge of the network's architecture. Although the Minos classifier's architecture was published by Naseem *et al.*, the training data and model were not made available. Therefore, we use a MINOS classifier trained by Cabrera-Arteaga *et al.* [7] as the target of our attack experiments.

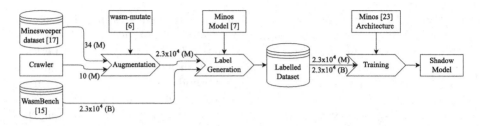

Fig. 1. Systematic overview of the training procedure for the substitute model. Malicious (M) samples are augmented to generate a balanced dataset. To generate labels, the target model is queried. The labelled benign (B) and malicious data is used to train the substitute model using 5-fold cross-validation.

3.1 Data Acquisition

The performance of the attack correlates with the quality of the local substitute model trained by the adversary. Therefore a comprehensive dataset of malicious and benign WebAssembly binaries is required to train a suitable substitute network. The original MINOS model was trained on a balanced dataset containing 300 samples [23]. The data preparation and training procedure for the substitute model is schematically visualized in Fig. 1 and described below in detail. To obtain benign samples, we used WasmBench[1], a WebAssembly dataset containing more than 23.000 real-world binaries published by Hilbig *et al.* as part of an empirical study [14]. We obtained 34 malicious samples from a dataset[2] published in the context of Minesweeper [17]. Additionally, we ran a crawler to increase the number of malware samples and gather up-to-date malware. By iterating over the Cisco Umbrella 1 Million list [11], we were able to download 187 WebAssembly binaries. Each domain on this list is visited by the crawler, which resides on any page for three seconds. By hooking a JavaScript function into each document load, we are able to dump any WebAssembly binary before it is executed. Considering that the malware may not reside on the homepage directly, the crawler additionally visits three randomly chosen internal links. Overall 40% of the crawled binaries resided on subdomains and were found either through accessing internal links or redirects. The Minesweeper [17] classifier categorized ten out of the 187 crawled binaries as being malicious. Even after combining the samples of public datasets with the results of our crawling campaign, the number of obtained malicious binaries is considerably lower than that of benign binaries. In order to compensate for this difference and additionally increase the number of samples, we utilize the Wasm-fuzzer wasm-mutate [6] as a diversifier. By utilizing wasm-mutate, one can generate a variety of different WebAssembly binaries that retain the original semantic. Mutation cores available in wasm-mutate enable semantic-preserving transformations. A sample function that performs the addition of two integers and two mutations of the function are shown in

---

[1] https://github.com/sola-st/WasmBench (Accessed 2023/01/31).
[2] https://github.com/vusec/minesweeper (Accessed 2023/01/31).

(a) Original function	(b) Mutated version of (a)	(c) Mutated version of (a)
| ```
(func (;0;)
  (param i32 i32)
  (result i32)
  local.get 0
  local.get 1
  i32.add
)
``` | ```
(func (;0;)
  (param i32 i32)
  (result i32)
  local.get 0
  i32.const 0
  i32.shl
  i32.const 0
  i32.add
  local.get 1
  i32.add
)
``` | ```
(func (;0;)
  (param i32 i32)
  (result i32)
  local.get 0
  local.get 1
  i32.add
  i32.const 0
  i32.shr_u
  local.get 0
  local.get 1
  i32.add
  i32.const 0
  i32.sub
  i32.and
)
``` |

Fig. 2. Wasm function performing the addition of two integers (a) and two semantic-preserving mutations (b),(c) of the original function using different seeds in `wasm-mutate` [6].

Fig. 2. Each mutation is generated using a different seed, allowing us to generate a larger variety of syntactically different binaries with identical semantics. To generate appropriate adversarial examples, a shadow model that is as similar to the target model as possible must be utilized. To achieve this, the internal labels assigned to the samples are only used for balancing and not used for training. Instead, the pre-trained MINOS network [7] is employed for label generation. After augmentation of the malicious samples, we obtain a dataset containing 2.3×10^4 malicious and 2.3×10^4 benign binaries that are used for training the substitute model.

3.2 Substitute Network Training

We use the architecture employed by MINOS for the substitute model because we assume a known architecture in the grey-box attack. The architecture of the CNN is shown in Fig. 3. Convolutional neural networks typically receive an image as the input for classification. The MINOS classifier requires the input to be a grey-scale image of size 100×100. To allow binaries of varying sizes to be represented as a fixed-dimensional image, the bytes are reshaped into the largest possible two-dimensional array with the same width and height. The remaining bytes are discarded. Initially, each byte of the binary corresponds to one pixel. However, the image is downscaled to a 100×100 image. A detailed description of the downsampling process is given in Sect. 3.3. The original model was trained using an 80% training and 20% testing split. However, we use 5-fold cross-validation for training. Hence five models are trained each on 80% of the dataset described in Sect. 3.1, while 20% of samples are withheld for validation. For the evaluation,
```

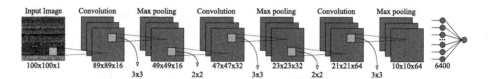

Fig. 3. Architectural overview of the MINOS classifier from Naseem *et al.* [23]. The CNN contains three convolution layers, three pooling layers, and one fully connected layer. The input image shows a `Wasm` binary that is transformed into a grey-scale image.

MINOS was trained with one epoch (M-1) to prevent overfitting, followed by 50 epochs (M-50), the same number as the target model. The area under the curve (AUC) and loss after the final epochs are reported in Table 1. Even after training the substitute network for only one epoch, the validation AUC reaches 99% with a validation loss of 0.14. After training for 50 epochs, the validation loss decreases to 0.04.

3.3   Attack Methodology

Performing an adversarial attack against an image-based classifier requires slight modifications of the original image to manipulate the generated response in the desired direction. The alterations are often transparent to the naked eye as they result in a small amount of noise added to the original image. However, in the case of binaries, slightly manipulating the value of a pixel, for example, changing a value from $0x2A$ to $0x2B$, changes the original instruction from `f32.load` to `f64.load` invalidating the binary. We require a procedure that allows us to manipulate certain areas of the binary without changing the behaviour. Using instrumentation, we can add, manipulate or remove instructions from the malware and provide areas inside the code section that can be utilized for the adversarial attack. While we are still unable to manipulate arbitrary pixels, adding specially crafted gadgets into the binary enables specific bytes to be utilized for the adversarial attack. Generating an adversarial example requires iterative manipulation of the target value in small increments. Hence, an area of bytes that

Table 1. Substitute network training evaluation after the last epoch for (a) one epoch (M-1) and (b) 50 epochs (M-50).

	(a) M-1						(b) M-50				
Fold	0	1	2	3	4	Fold	0	1	2	3	4
AUC	0.96	0.95	0.96	0.96	0.94	AUC	1.00	1.00	1.00	1.00	1.00
Val. AUC	0.99	0.99	0.99	0.99	0.99	Val. AUC	1.00	1.00	1.00	1.00	1.00
Loss	0.29	0.30	0.29	0.29	0.34	Loss	0.03	0.03	0.04	0.03	0.03
Val. Loss	0.13	0.14	0.15	0.15	0.14	Val. Loss	0.05	0.05	0.03	0.04	0.04

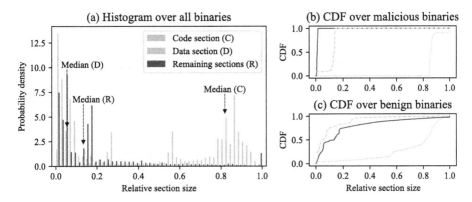

Fig. 4. Histogram of relative section size (a) for code section, data section and all remaining sections for all binaries as described in Sect. 3.1. Cumulative density for relative section size for all malicious binaries (b) and benign binaries (c).

are arbitrarily manipulable is ideal. Each WebAssembly binary is split into several sections, each with a different purpose. As shown in Fig. 4a, the code section represents, in most cases, the largest section inside both malicious and benign binaries that were analyzed. When separately evaluating the section distribution for malicious and benign binaries (cf. Fig. 4b and Fig. 4c), it is apparent that in both cases, the code section remains the largest section. The code section contains all functions with their instructions, whereas the data section represents a linear array of memory accessible through instructions in the code section. While an attack against the data section is also possible by extending the size of the linear memory and using this area for crafting the attack, we chose to target the code section as it represents the largest section of the binaries. An overview of our attack methodology is given in Fig. 5. Each step is described in detail below.

Semantic-Preserving Gadgets. To enable manipulation inside the code section, we require an instruction that has a number of bytes that are freely choosable. In particular, instructions that load constants onto the stack cause specific values to be present inside the code section. Hence, constructing a gadget that loads an arbitrary constant onto the stack and removes it allows a number of bytes to be arbitrarily chosen. Additionally, it can be inserted anywhere into the control flow because, after the gadget's execution, the stack will be in the same state as before. WebAssembly allows four number types to be pushed onto the stack as constants - 32 and 64 bit variants of integers and floats. We opt to use 64 bit constants, as the ratio between the number of bytes that are available for the adversarial attack and the number of bytes required for the overall gadget is higher. Generally, both integers and floats can work. However, WebAssembly encodes all integers using the *LEB128* variable-length encoding in either the signed or unsigned variant. Compared to the encoding utilized for floating point values, *IEEE-754* [2], the integer encoding enforces a number of

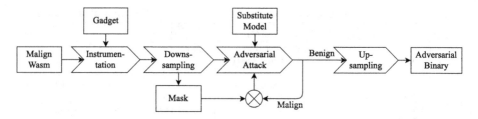

Fig. 5. Schematic overview of the attack methodology. A malicious binary is instrumented to add the gadgets used for carrying the adversarial payload. After downsampling, the adversarial attack is performed against the substitute model. To recreate the original binary, we upsample the adversarial image and recreate the original binary.

restrictions on the bytes representing the integer. *IEEE-754*, on the other hand, allows all bytes to assume all possible values. Hence we use 64 bit floating point constants to craft the attack. The `f64.const x:f64` instruction can be used to push the 64 bit floating point number x onto the stack. We initialize the constant to $0x80808080$ to allow both positive and negative perturbations. To ensure that the functionality of the target binary is not modified, the value must be removed from the stack before normal execution resumes. We demonstrate two gadgets that can be inserted after arbitrary instructions, as the execution of the gadget only changes the contents of the stack temporarily. A *size-efficient gadget* (SE) is shown in Fig. 6a. After the constant is pushed onto the stack, it is immediately removed again using the `drop` instruction. Each inserted gadget of this type increases the size of the binary by ten bytes, out of which the adversarial attack can utilize eight bytes (compare Fig. 6b). Hence, only 20% of the size overhead is attributed to bytes that cannot be manipulated during the attack phase.

Due to the low complexity of the size-efficient gadget, it is easy to discern that the two instructions will retain the program's semantics. However, optimizers such as `wasm-opt` [38] can remove all gadgets of this type from the binary. Note that using an optimizer before classifying the binary is not part of the MINOS framework [23] because it would counteract the high efficiency of the detection system. Nevertheless, we are able to craft a gadget that is not removed by `wasm-opt`, even when using its most aggressive optimization setting. This resilience, however, is only made possible by increasing the gadget's complexity. The composition of our *optimizer-resistant gadget* (OR) is shown in Fig. 6c and the binary representation in Fig. 6d. The basic idea remains unchanged; we still load a constant onto the stack, thus introducing a value that can be manipulated during the attack phase. However, instead of directly loading the value onto the stack and dropping it, we use it as the increasing constant for a loop counter. However, as the value can be an arbitrary float value, i.e. negative and positive, we divide it by itself to have a known value, i.e. one. We then check whether this

(a) Size-efficient gadget	(b) Binary representation of (a)
`f64.const 0x0` `drop`	`0x44,0x0,0x0,0x0,0x0,0x0,0x0,0x0,0x0` `0x1a`

(c) Optimizer-res. gadget	(d) Binary representation of (c)

`(local $UID1 f64)`
`(loop $UID2`
`  local.get $UID1`
`  f64.const 0x0`
`  f64.add`
`  local.tee $UID1`
`  local.get $UID1`
`  f64.div`
`  f64.const 42`
`  f64.gt`
`  br_if $UID2`
`)`
	
	`0x03,0x40`
	`0x20,0x02`
	`0x44,0x0,0x0,0x0,0x0,0x0,0x0,0x0,0x0`
	`0xa0`
	`0x22,0x02`
	`0x20,0x02`
	`0xa3`
	`0x44,0x0,0x0,0x0,0x0,0x0,0x0,0x45,0x40`
	`0x64`
	`0x0d,0x00`
	`0x0b`

Fig. 6. Size-efficient (a) and optimizer-resistant (c) gadget and their binary representation (b, d). Bytes that can be manipulated during the adversarial attack are highlighted in blue. (Color figure online)

new value is less than some constant, i.e. 42, which is always true, and break the loop. While it is intuitively understandable that this loop will never be executed more than once, it is not easily determined by an algorithm since loops are difficult to analyze. While this gadget survives optimization passes, only eight out of 32 bytes can be utilized for the adversarial attack. Gadgets are inserted into the code section at randomly drawn insertion points with a predetermined frequency. The relation between the number of inserted gadgets and the success rate of the attack is evaluated in Sect. 4.1. In Sect. 4.2, we evaluate the execution speed of both gadgets in relation to the number of gadgets inserted into the binary. Insertion of either gadget into the target binary can be performed once per binary before distribution and requires linear time in the size of the binary, making the instrumentation efficient.

Downsampling. A given binary can be of any size between a few kilobytes and many megabytes. Hence, the authors of Minos [23] downsample each binary into an image of fixed dimensionality, i.e. 100×100 pixels (Fig. 3). As our shadow model utilizes the same architecture, it also requires an input image of that size. However, as we need to keep track of the positions that allow for a change within the instrumented binary, i.e. the constants within our gadgets, we use a custom downsampling algorithm for crafting the attacks. Yet, at inference time, the original downsampling method is used. At first, we transform the sequence of bytes b from the binary into a squared image with a dimension of $\lfloor \sqrt{|b|} \rfloor$. Hence, a few bytes at the end are discarded. From this squared image, we combine as many pixels as needed in order to downsample the image to 100×100 pixels. For this purpose, we calculate the mean of a group of pixels, which then become a single pixel. To keep track of what pixels contain a byte that is used for the adversarial attack, we maintain a mask M_1. The mask has the same dimensionality as the image and marks all positions that contain editable values. To easily revert the downsampling when restoring the binary, we store the coordinates of the original group of pixels for each downsampled pixel.

Adversarial Attack. After downsampling, the image x is perturbed iteratively until our shadow model misclassifies the image as benign using the method proposed by Carlini & Wagner [8]. However, instead of optimizing for a fixed number of iterations, we keep iterating until the shadow model prediction reaches a threshold τ. Experimentally we determined $\tau = 10^{-13}$. However, we also terminate the optimization after 1×10^4 iterations. During our experiments, we found that the lower the threshold for the prediction score is, the higher the chance that an original model will share the classification of the shadow model. In order to only perturb pixels related to the gadgets, we multiply the mask M_1 that was saved during downsampling before adding the perturbation δ_x to the sample. Given the model θ, a normalization $|\cdot|$ and the constant ϵ, the perturbation of the input under the objective function $\mathcal{L}$ is given as:

$$x = x + M_1 \cdot \epsilon \cdot \left| \frac{\mathrm{d}}{\mathrm{d}x} \mathcal{L}(\theta(x), 0) \right|$$

In our experiments, we chose $\epsilon = 0.05$ and $\mathcal{L}$ as binary cross-entropy [5]. We derive the change needed for the input x within the normalization term so that the prediction $\theta(x)$ gets closer to zero, i.e. benign. However, instead of adding the whole perturbation to x, only a small factor is added. This can be compared to the learning rate in classical machine learning. As we cannot perturb the whole input image but rather just the constants within the gadgets, our crafted mask is multiplied before the summation. As the mask has zeros on all non-editable pixels, i.e. the original code of the binary, and a one wherever there is at least a single gadget, the perturbation is only applied to pixels that relate to gadgets.

Upsampling. The result of the adversarial attack is a perturbed image x^* where the perturbation is only applied to the pixels that initially belonged to at least a single gadget. Those changes must now be mapped back to the original binary. For the perturbed image, we look at every pixel that belonged to at least one gadget. If such a pixel is found, we retrieve the corresponding group of pixels $\mathcal{G}$. To correctly update $\mathcal{G}$, the bytes belonging to an adversarial payload need to be modified s.t. the mean value of $\mathcal{G}$ equals the corresponding pixel value of x^*. Given the sum of the pixel values $\sum_{p \in \mathcal{G}} p$, the number of pixels $|\mathcal{G}|$ and the target pixel p^* the update factor f_{adv} can be derived using the following equation:

$$f_{adv} = p^* \cdot |\mathcal{G}| - \sum_{p \in \mathcal{G}} p$$

To apply the factor f_{adv} to the adversarial payload, we create a mask M_2 that has a one at every editable position within $\mathcal{G}$. $\overline{M_2}$ contains the same values as M_2 but flipped, s.t., ones become zeros and vice versa. We can update the group of pixels using the following equation:

$$\mathcal{G}_{adv} = \begin{cases} M_2 \dfrac{f_{adv}}{\sum M_2} + \overline{M_2}\mathcal{G} & \text{if } \sum M_2 \geq 1 \\ \mathcal{G} & \text{otherwise} \end{cases}$$

The left term of the addition in the first case replaces all the editable pixels within the image with a shared factor. The second term adds the original values. This way, the new mean value of $\mathcal{G}_{adv}$ equals the target value of the downsampled image. In case there are no gadgets in the particular group, i.e. $\sum M_2 = 0$, $\mathcal{G}$ is simply copied. After the termination of the adversarial attack, the image is flattened into a byte array b_{adv}, and the bytes that were cropped during downsampling are appended again.

Possible Countermeasures. In Sect. 4.1, we show that MINOS [23] is susceptible to the presented adversarial attack. However, it is essential to also discuss possible improvements that could prevent such adversarial attacks and aid in hardening the detection framework. The option to remove semantic-preserving gadgets using an optimizer was already discussed in Sect. 3. While an additional optimization step prevents an adversary from relying on the size-efficient gadget, the more complex optimization-resistant gadget still allows effective adversarial attacks. Machine learning models can be directly hardened against adversarial attacks using, for example, defensive distillation [26], which is a technique where the class probability vectors of a trained DNN are used to train another DNN of the same dimensionality. As the name suggests, defensive distillation is derived from the concept of distillation [3], where one trained DNN is used to train a smaller DNN without losing accuracy. Another promising method for hardening models against adversarial attacks is presented by Goodfellow et al. [13]. They create adversarial examples and use them as training data for their model. However, the presented countermeasures were shown not to be effective against a thoughtful attacker [36].

4   Evaluation

4.1   Gadget Effectiveness

Using our corpus of malicious samples (Sect. 3.1), we evaluate the effectiveness of our attack by creating adversarial examples for each binary. We consider the insertion density d as the relative frequency of occurrence of our gadget, s.t. for a given density $d \in [0, 1]$, for every 1000 instructions $d \cdot 1000$ gadgets are added. Figure 7 shows the misclassification rates of binaries with the size-efficient gadget (Fig. 7a) and the optimization-resistant gadget (Fig. 7b) against the MINOS

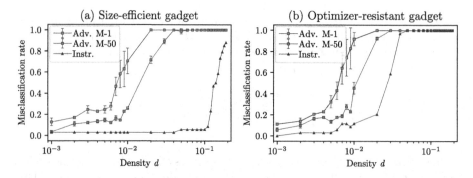

Fig. 7. MINOS misclassification rate of binaries with size-efficient gadgets (a) and optimizer-resistant gadgets (b) against the pre-trained MINOS [23] classifier by Cabrera-Arteaga *et al.* [7]. Each plot depicts the misclassification rate of the original binary (Original), the instrumented binary *without* adversarial payload (Instr.), and the misclassification rate of the binaries *with* adversarial payload derived using Minos trained for one epoch (Adv. M-1) and for 50 epochs (Adv. M-50). The adversarial misclassification rates are average over all five folds. The error bars depict the standard deviation.

classifier [23] trained by Cabrera-Arteaga et *al.* [7]. To the best of our knowledge, MINOS is the only WebAssembly malware classifier that utilizes machine learning to classify malware directly on a representation of the binary itself. To evaluate the effectiveness of our adversarial payloads at invoking misclassifications, we plot the misclassification rates for the original binary, the instrumented binary without adversarial payload and the adversarially crafted binaries. The original binaries are unaffected by the gadget density and never result in misclassification. For instrumented binaries without adversarial payloads, it becomes apparent that after a sufficiently large number of insertions, the classifier cannot detect the malicious binary even without the adversarial attack. Figure 8b shows the size increase of the binary through the addition of our gadgets. For each gadget, the misclassification rates of the instrumented binaries start to increase significantly at a size of roughly 1.5× the original binary. Considering that the larger the binary gets, the higher the compression rates and information loss are during downsampling, an increase in misclassification rates that correlates with a size increase can occur. Due to the difference in the number of added bytes per gadget, the misclassification rate for the larger optimization-resistant gadget increases at lower densities. However, for both gadgets, one can observe that the adversarially crafted binaries consistently outperform the binaries that are only instrumented, causing higher misclassification rates at lower densities. Additionally, adversarial payloads generated using the substitute models trained for one epoch consistently cause higher misclassifications at lower densities than payloads generated using the models trained for 50 epochs. To further evaluate the misclassification caused by instrumenting the malicious binary, we additionally instrumented 50 randomly selected benign binaries with the optimizer-resistant

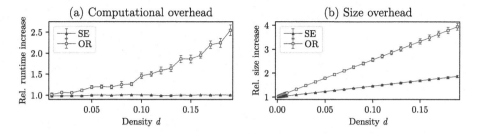

Fig. 8. Correlation between the insertion density and the relative increase in execution time (a) and size (b). Both the size-efficient gadget (SE) and the optimization-resistant gadget (OR) are evaluated. The x-axis represents the density of the gadgets, while the y-axis represents the relative execution time compared to the baseline (no gadget insertion) (a) and the relative increase of the binary's size in bytes (b). The average over the evaluated binaries is plotted, and the error bars represent the standard deviation.

gadget that caused higher misclassification rates. At densities of both 0.1 and 0.01, the classifier correctly identified all evaluated benign binaries as benign, suggesting a tendency of the classifier to classify samples as benign. To evaluate the effectiveness of our method, we additionally generated adversarial payloads for the benign binaries that caused the substitute model to misclassify the binary as malicious. Using the substitute model trained for one epoch, we were able to successfully cause the target classifier to misclassify, on average, 77% of the binaries over all folds at a density of 0.1. Overall, at a density of 0.02, both gadgets are shown to be successful in evading the target classifier for at least 70% of evaluated malicious binaries, while the misclassification rates for the instrumented binary without the adversarial payload are at or below 20%, highlighting the effectiveness of our approach.

4.2 Performance Analysis

To quantify the gadget's impact on the runtime of instrumented binaries, we measured the execution time in relation to the gadget density. This correlation is illustrated in Fig. 8a. We utilized a WebAssembly hashing library [24] and performed 5×10^5 rounds of SHA-256 hashing. A baseline was established by measuring the execution time without inserting the gadgets. The execution time of both gadgets is shown in relation to the baseline. The insertion of the size-efficient gadget only results in a small constant increase in execution time, suggesting that the inserted gadget is not executed. WebAssembly is compiled using an ahead-of-time compiler, which includes optimization of the code. As the size-efficient gadget neither changes the data flow nor the control flow, the compiler likely identifies and removes those instructions during compilation. However, similar to `wasm-opt` [38], this optimizer cannot detect the optimization-resistant gadget. As a result, the execution time increases linearly in the number of inserted gadgets. However, considering that a density of 0.02 is enough to trick the target classifier, the increase in runtime is reasonable.

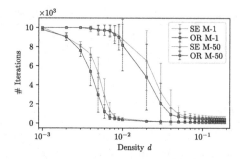

Fig. 9. Average number of iterations (y-axis) required to achieve a confidence of $1\times^{-12}$ for a given gadget density (x-axis). Both the size-efficient gadget (SE) and the optimizer-resistant gadget (OR) are evaluated on the substitute model trained for one epoch (M-1) and 50 epochs (M-50). The error bars show the standard deviation.

Additionally, we evaluated the requirements for generating an adversarial example, which heavily depends on the gadget density. The number of iterations required to achieve a confidence of less than 1×10^{-13} within the shadow model was measured as a function of the chosen gadget density. The results are depicted in Fig. 9, which displays the average number of iterations required during the adversarial example generation over the applied gadget density. As both gadget types hold the same number of bytes utilized for the adversarial payload, they require a similar number of iterations to reach the confidence level. The adversarial training optimization loop was run for a maximum of 1×10^4 iterations. Overall, the lower the chosen density, the more iterations are required to reach the target confidence, as fewer bytes are available for adversarial crafting. While the adversarial examples crafted using the substitute model trained for one epoch outperform the adversarial examples crafted using the model trained for 50 epochs, the adversarial example reaches the target confidence with fewer iterations on the model trained for 50 epochs. The execution time of a single iteration is 9.84 ms on an AMD Ryzen 9 7950X 16-Core Processor, which renders the attack feasible. Note that this optimization needs to only be performed once per malware. However, an attacker could potentially exploit the low cost of generating new adversarial examples by regularly distributing new binaries to website visitors.

5   Related Work

The use of machine learning-based classifiers for detecting malware has been shown to be fast and effective in identifying binaries as malicious or benign. However, the robustness of these classifiers against adversarial inputs is often limited.Cite As more machine learning-based classifiers are utilized for detecting malware, malicious actors who want to distribute their malware have a high incentive to utilize evasion techniques to prevent detection. Especially for Windows Portable Executables (PEs), a number of classifiers and evasions exist.

Existing adversarial evasions on classifiers that utilize a gray-scale image representation of the target binary [16] rely on FSGM [13] or Carlini & Wagner [8], to generate a perturbation vector for the image [16,20,28]. However, in contrast to our attack, Liu *et al.* [20] directly apply the perturbation to the image representation of the binary. While they show a successful attack against the classifier, the generated adversarial example is not a valid binary anymore, rendering their evasion ineffective. Khormali *et al.* [16] generate the adversarial example and append the adversarial payload to the end of the file or at the end of a section. This ensures that the adversarial example is added into nonexecutable areas, and hence the original functionality remains. While this enables the addition of the adversarial payload into the malicious binary, a sophisticated defender can easily remove the payload by statically identifying unused bytes and masking them before classification, as they should have no impact on the classification performance. Using our attack methodology, the adversarial payload is placed inside the code section and directly baked into the control flow of the target binary, preventing a defender from easily removing the payload. Additionally, we have presented the optimization-resistant gadget that cannot be generally removed using an optimization pass. Evasions against other network architectures that directly consider the sequence of bytes from Windows PE filescite generally insert adversarial payloads in unused bytes between sections [18,29,33], in a new section [18] or at the end of the file [9,29]. While these approaches generate executable binaries, it is rather easy to circumvent for a slightly more sophisticated detection model, e.g. one that first removes unused bytes or truncates sections or files. Either of our proposed gadgets is inserted directly into the instructions so that more sophisticated static analysis techniques, such as data flow and control flow analysis, are required to detect them fully. However, there are also numerous adversarial attacks against classifiers that classify a binary on more sophisticated features than just an image from its raw binary data, e.g. based on extracted features such as control flow, data flow, API calls, libraries, or dynamic features [10,19]. While the general procedure for generating the perturbation vector is similar, the application to the binary relies on transforming the target in a way that the corresponding features change. The interested reader is referred to Ling *et al.* [19], who provide an in-depth evaluation of different evasion techniques against Windows PE malware. Cabrera-Arteaga *et al.* [7] proposed a malware evasion system against Wasm malware detectors and, in particular, MINOS. However, their system relies on obfuscation to bypass detection frameworks, and they do not utilize adversarial attacks.

6   Conclusion

In this paper, we introduced a novel technique for placing adversarial payloads directly into the instruction stream using binary instrumentation to bypass machine learning-based malware detectors. We have demonstrated the effectiveness of our technique by crafting a grey-box adversarial attack against MINOS [23], a lightweight cryptojacking detection framework for WebAssembly presented at NDSS 2021. To place payloads inside the code section of the binary, we have introduced two semantic-preserving gadgets for `Wasm` binaries with a focus on size-efficiency and optimization-resistance, respectively. We have collected an extensive dataset with both benign and malicious binaries by utilizing two existing benchmark datasets [14,17] as well as results from a crawling campaign of one million websites from the Cisco Umbrella list [11]. To populate this dataset, we used `wasm-mutate` [6] to generate augmented binaries. Every sample was then assigned a label by querying the target model, i.e. MINOS [23] provided by Cabrera-Arteaga *et al.* [7]. All samples with their corresponding label were then used to train a substitute model of our targeted model. The challenge of creating a functional adversarial example inside a binary without altering the semantics was met by carefully inserting novel semantic-preserving gadgets. These gadgets can be injected freely into the code section of a `Wasm` binary without changing the semantics using binary instrumentation. Each gadget contains a number of bytes that carry the adversarial payload and can be manipulated freely during the attack phase. By attacking our substitute model, we successfully craft functional adversarial examples for cryptojacking binaries. Using an insertion density of 0.02 and the better-performing substitute network trained for one epoch (M-1), we are able to cause the target detector to misclassify all of the evaluated malicious binaries, demonstrating the effectiveness of our attack. Additionally, we show that our size-efficient gadget is removed during compilation resulting in only a negligible runtime overhead. The optimizer-resistant gadget, by design, is not removed before execution and thus leads to a linear overhead in the density. However, as a small insertion density of 0.02 is sufficient in bypassing the classifier, the execution time is only increased by roughly 10%. To prevent such attacks, we addressed typical countermeasures; However, as discussed by Tramèr *et al.* [36], as long as the adversary is able to manipulate features used by a classifier, the threat of adversarial attacks cannot be fully mitigated. The success of our grey-box adversarial attack on MINOS highlights the need for continued research and improvement of defences against adversarial attacks on machine learning-based malware detection frameworks.

Acknowledgements. We thank the reviewers and our shepherd for their helpful comments and suggestions. This work has been supported by ERDF through the EMSIK project and by BMBF through the PeT-HMR project.

References

1. WebAssembly Core Specification. https://www.w3.org/TR/wasm-core-2/
2. Ieee standard for floating-point arithmetic. IEEE Std 754–2019 (Revision of IEEE 754–2008) (2019)
3. Ba, J., Caruana, R.: Do deep nets really need to be deep? In: Advances in Neural Information Processing Systems, vol. 27 (2014)
4. Biggio, B., Corona, I., Maiorca, D., Nelson, B., Šrndić, N., Laskov, P., Giacinto, G., Roli, F.: Evasion attacks against machine learning at test time. In: Blockeel, H., Kersting, K., Nijssen, S., Železný, F. (eds.) ECML PKDD 2013. LNCS (LNAI), vol. 8190, pp. 387–402. Springer, Heidelberg (2013). https://doi.org/10.1007/978-3-642-40994-3_25

5. de Boer, P., Kroese, D.P., Mannor, S., Rubinstein, R.Y.: A tutorial on the cross-entropy method. Ann. Oper. Res. **134**(1) (2005)
6. Bytecodealliance: wasm-mutate (2023). https://github.com/bytecodealliance/wasm-tools/tree/main/crates/wasm-mutate. Accessed: 2023
7. Cabrera-Arteaga, J., Monperrus, M., Toady, T., Baudry, B.: Webassembly diversification for malware evasion (2022)
8. Carlini, N., Wagner, D.A.: Towards evaluating the robustness of neural networks. In: 2017 IEEE Symposium on Security and Privacy, SP 2017 (2017)
9. Chen, B., Ren, Z., Yu, C., Hussain, I., Liu, J.: Adversarial examples for CNN-based malware detectors. IEEE Access 7 (2019)
10. Chen, L., Ye, Y., Bourlai, T.: Adversarial machine learning in malware detection: arms race between evasion attack and defense. In: European Intelligence and Security Informatics Conference, EISIC 2017 (2017)
11. Dan Hubbard: Cisco Umbrella 1 Million (2021). https://umbrella.cisco.com/blog/cisco-umbrella-1-million. Accessed: 2023
12. Eskandari, S., Leoutsarakos, A., Mursch, T., Clark, J.: A first look at browser-based cryptojacking. In: 2018 IEEE European Symposium on Security and Privacy Workshops, EuroS&P Workshops 2018 (2018)
13. Goodfellow, I., Shlens, J., Szegedy, C.: Explaining and harnessing adversarial examples (2014)
14. Hilbig, A., Lehmann, D., Pradel, M.: An empirical study of real-world webassembly binaries: security, languages, use cases. In: WWW '21: The Web Conference 2021. ACM/IW3C2 (2021)
15. Kharraz, A., et al.: Outguard: detecting in-browser covert cryptocurrency mining in the wild. In: The World Wide Web Conference, WWW 2019 (2019)
16. Khormali, A., Abusnaina, A., Chen, S., Nyang, D., Mohaisen, A.: Copycat: practical adversarial attacks on visualization-based malware detection (2019)
17. Konoth, R.K., Vineti, E., Moonsamy, V., Lindorfer, M., Kruegel, C., Bos, H., Vigna, G.: Minesweeper: an in-depth look into drive-by cryptocurrency mining and its defense. In: Proceedings of the 2018 ACM SIGSAC Conference on Computer and Communications Security, CCS (2018)
18. Kreuk, F., Barak, A., Aviv-Reuven, S., Baruch, M., Pinkas, B., Keshet, J.: Deceiving end-to-end deep learning malware detectors using adversarial examples (2018)
19. Ling, X., et al.: Adversarial attacks against windows PE malware detection: a survey of the state-of-the-art. CoRR (2021)
20. Liu, X., Zhang, J., Lin, Y., Li, H.: ATMPA: attacking machine learning-based malware visualization detection methods via adversarial examples. In: Proceedings of the International Symposium on Quality of Service, IWQoS 2019 (2019)
21. Musch, M., Wressnegger, C., Johns, M., Rieck, K.: New kid on the web: a study on the prevalence of WebAssembly in the wild. In: Perdisci, R., Maurice, C., Giacinto, G., Almgren, M. (eds.) DIMVA 2019. LNCS, vol. 11543, pp. 23–42. Springer, Cham (2019). https://doi.org/10.1007/978-3-030-22038-9_2
22. Nakamoto, S.: Bitcoin: a peer-to-peer electronic cash system. Decentralized business review, p. 21260 (2008)
23. Naseem, F.N., Aris, A., Babun, L., Tekiner, E., Uluagac, A.S.: MINOS: a lightweight real-time cryptojacking detection system. In: 28th Annual Network and Distributed System Security Symposium, NDSS 2021 (2021)
24. Noh, J.: WebAssembly Works (2008). https://github.com/Snack-X/wasm-works. Accessed: 2023
25. Papadopoulos, P., Ilia, P., Markatos, E.P.: Truth in web mining: Measuring the profitability and cost of cryptominers as a web monetization model (2018)

26. Papernot, N., McDaniel, P., Wu, X., Jha, S., Swami, A.: Distillation as a defense to adversarial perturbations against deep neural networks. In: Proceedings - 2016 IEEE Symposium on Security and Privacy, SP 2016 (2016)
27. Papernot, N., McDaniel, P.D., Goodfellow, I.J., Jha, S., Celik, Z.B., Swami, A.: Practical black-box attacks against machine learning. In: Proceedings of the 2017 ACM on Asia Conference on Computer and Communications Security, AsiaCCS 2017 (2017)
28. Park, D., Khan, H., Yener, B.: Generation & evaluation of adversarial examples for malware obfuscation. In: 18th IEEE International Conference On Machine Learning And Applications, ICMLA 2019 (2019)
29. Qiao, Y., Zhang, W., Tian, Z., Yang, L.T., Liu, Y., Alazab, M.: Adversarial malware sample generation method based on the prototype of deep learning detector. Comput. Secur. **119** (2022)
30. Rodriguez, J.D.P., Posegga, J.: RAPID: resource and api-based detection against in-browser miners. In: Proceedings of the 34th Annual Computer Security Applications Conference, ACSAC 2018 (2018)
31. Romano, A., Zheng, Y., Wang, W.: Minerray: semantics-aware analysis for ever-evolving cryptojacking detection. In: 35th IEEE/ACM International Conference on Automated Software Engineering, ASE 2020 (2020)
32. Russo, M., Srndic, N., Laskov, P.: Detection of illicit cryptomining using network metadata. EURASIP J. Inf. Secur. **2021**(1) (2021)
33. Suciu, O., Coull, S.E., Johns, J.: Exploring adversarial examples in malware detection. In: 2019 IEEE Security and Privacy Workshops, SP Workshops 2019 (2019)
34. Szegedy, C., et al.: Intriguing properties of neural networks. In: 2nd International Conference on Learning Representations, ICLR 2014 (2014)
35. Tekiner, E., Acar, A., Uluagac, A.S., Kirda, E., Selçuk, A.A.: Sok: Cryptojacking malware. In: IEEE European Symposium on Security and Privacy, EuroS&P (2021)
36. Tramèr, F., Carlini, N., Brendel, W., Madry, A.: On adaptive attacks to adversarial example defenses. In: Advances in Neural Information Processing Systems 33: Annual Conference on Neural Information Processing Systems 2020, NeurIPS 2020 (2020)
37. Wang, W., Ferrell, B., Xu, X., Hamlen, K.W., Hao, S.: SEISMIC: SEcure in-lined script monitors for interrupting cryptojacks. In: Lopez, J., Zhou, J., Soriano, M. (eds.) ESORICS 2018. LNCS, vol. 11099, pp. 122–142. Springer, Cham (2018). https://doi.org/10.1007/978-3-319-98989-1_7
38. WebAssembly: Binaryen (2022). https://github.com/WebAssembly/binaryen. Accessed: 2023
39. Zhang, J., Jiang, X.: Adversarial examples: Opportunities and challenges (2018)

Honey, I Chunked the Passwords: Generating Semantic Honeywords Resistant to Targeted Attacks Using Pre-trained Language Models

Fangyi Yu$^{(\boxtimes)}$ and Miguel Vargas Martin$^{(\boxtimes)}$

Ontario Tech University, Oshawa, ON L1G 0C5, Canada
{fangyi.yu,miguel.martin}@ontariotechu.ca

Abstract. Honeywords are fictitious passwords inserted into databases in order to identify password breaches. The major challenge is producing honeywords that are difficult to distinguish from real passwords. Although the generation of honeywords has been widely investigated in the past, the majority of existing research assumes attackers have no knowledge of the users. These honeyword generating techniques (HGTs) may utterly fail if attackers exploit users' personal identifiable information (PII) and the real passwords include users' PII. The literature has demonstrated that password guessing is more effective when focusing on each of the chunks that compose a password (e.g., "P@ssword123" contains two chunks: "P@ssword" and "123") and it has been suggested that, when available, PII should be used to generate honeywords. We thus leverage these findings to base our HGT method on any possible PII contained within passwords, and introduce a new, and more robust than its literature counterparts, method to generate honeywords, which consists of generating honeywords with GPT-3 using the semantic chunks of their corresponding real passwords.

Furthermore, we propose a new metric, HWSimilarity, to evaluate the capability of HGTs. HWSimilarity is a pre-trained language model-based similarity metric that considers the semantic meaning of passwords when measuring the indistinguishability of honeywords and their counterparts. Comparing our chunk-level GPT-3 HGT to two state-of-the-art HGTs and using GPT-3 alone, we show that our HGT can generate honeywords that are more indistinguishable than its counterparts.

Keywords: authentication · chunking · honeywords · natural language processing · language models

1 Introduction

Passwords have dominated the authentication system for decades, despite their security flaws compared to competing techniques such as cognitive authentication [12], biometrics [20] and tokens [22]. Their irreplaceability is primarily due to their incomparable deployability and usability [3]. However, current password-based authentication systems store sensitive password files that make them ideal

© The Author(s), under exclusive license to Springer Nature Switzerland AG 2023
D. Gruss et al. (Eds.): DIMVA 2023, LNCS 13959, pp. 89–108, 2023.
https://doi.org/10.1007/978-3-031-35504-2_5

targets for attackers because if successfully obtained and cracked (recovering the hashed passwords' plain-text representations), an adversary may impersonate registered users in an undetectable fashion [26]. Numerous prestigious online services have been infiltrated, for example, Yahoo!, RockYou, Zynga, resulting in the exposure of millions of credentials. Unfortunately, there is often a large delay between a credential database's breach and its detection; estimates place the average latency at 287 days [1]. The resulting window of vulnerability enables attackers to crack passwords offline and use them directly to extract value or sell them via illicit forums profiting with stolen credentials [25]. Normally, the longer it takes to detect and remediate a data breach, the more expensive it is [1]. As a result, it is vital to have active, timely password-breach detection systems in place to allow immediate counter-actions.

One way to reduce the cost of password breaches is to make offline guessing harder [5]. However, this method has major disadvantages, such as low scalability or a need for large modifications to the server-side and client-side authentication systems, which prevent the community from implementing them. Another promising approach is to shorten the latency between password breaches and detection. Juels and Rivest suggest the use of honeywords as a potential method for efficiently detecting password leaks [13]. According to their proposal, a website could store decoy passwords, called honeywords, alongside real passwords in its credential database, so that even if an attacker steals and reverts the password file containing the users' hashed passwords, they must still choose a real password from a set of k distinct *sweetwords*, where a real password and its associated honeywords are referred to as sweetwords. The attacker's use of a honeyword could cause the website to become aware of the breach. Notably, honeywords are only beneficial if they are difficult to distinguish from real-world passwords; otherwise, a knowledgeable attacker may be able to recognize them and compromise their security. Thus, when implementing this security feature into current authentication systems, the honeyword generating process is critical.

1.1   Honeywords for Targeted Attacks

The biggest challenge of designing a HGT is to generate honeywords that are resistant to targeted attacks [28]. For targeted attacks, attackers exploit users' PII to guess passwords, which increases the likelihood of users' accounts being compromised. This is a critical problem because numerous PII and passwords become widely accessible as a result of ongoing data breaches [1] and people are used to create easy-to-remember passwords using their names, birthdays, and their variants [28]. Once an attacker obtains users' PII, and if only one sweetword in a user's sweetword list contains the user's PII, it is highly likely that this sweetword is the real password and others are fake. For example, for a sweetword list "*gaby1124, abg71993, australiaisno#1, 10L026378, noviembre9101, Elena1986@327, cken22305*" which are generated using a made-up password "*Elena1986@327*" (suppose this is the real password) and the HGT proposed by Dionysiou et al. [9]. In this case, if the attacker has no information about the user, it will be difficult to determine which of the seven sweetwords is the real password, since all of the honeywords are from data breaches and are legitimate passwords belonging to other users. However, if the attacker knows the user's first name is *"Elena"*, it is quite straightforward to deduce that *"Elena1986@327"* is this user's real password and all the others are fake.

Table 1. Data breaches containing PII and passwords in the past five years

Dataset	Number of Items	Year	Type of PII breached
Neiman Marcus	4,800,000	2021	Name, Encrypted Password, Security questions, Financial information
CAM4	10,880,000,000	2020	Name, Email, Encrypted Password, Chat transcripts, IP, Payment logs
Canva	137,000,000	2019	Name, Email, Encrypted Password
Quora	100,000,000	2018	Name, Email, Encrypted Password, Questions and answers posted
Yahoo	3,000,000,000	2017	Name, Email, Encrypted Password, DoB, Security question and answer

Following the introduction of the honeywords security mechanism by Juels and Rivest [13], the academic community has been actively exploring the technique. However, to our knowledge, only Wang et al. [29] concentrated on the production of honeywords in a targeted manner. All other works make the invalid assumption that attackers have no knowledge about the users. Each year, as demonstrated in Table 1, billions of password datasets including PII are leaked. Attackers might use the PII to determine which sweetword is the real password. If none of the sweetwords include PII existing in the password breach, the attackers may still create a knowledge map for each user by searching their information purposefully through social media and search engines using the known PII exposed in data breaches. This is especially a concern if the user is a public figure. Compromised accounts may have substantial financial, political, and societal consequences.

1.2   Related Work

Numerous studies have been conducted on the non-targeted honeyword generation method. The majority of these HGTs fall into two categories: chaffing-by-tweaking and chaffing-with-a-password-model. Chaffing-by-tweaking is mostly based on the substitution of random letters, digits, and symbols. For instance, given the real password *"deshaun96"*, we could get honeywords *"deshaun87, deshAUn66, DesHaun56"* via tweaking. However, as Wang et al. [26] demonstrate, this strategy is indeed vulnerable. While honeywords generated using the chaffing-with-a-password-model approach are more resistant to attacks, they do have certain drawbacks. Bojinov et al. [2] proposed *Kamouflage*, which first tokenizes the user's real passwords into a collection of tokens, and then substitutes each token with a random one that matches the token's type. For instance, *"jones34monkey"* is tokenized as *"$l_5d_2l_6$"* (a five-letter word followed by two digits and a six-letter word), indicating that some possible honeywords are *"apple10laptop, tired93braces, hills28highly"*. This technique, as outlined in [9], demands considerable modifications on the client-side authentication system, which has a significant impact on usability. Additionally, it is incapable of generating honeywords of varying length or structure, thus limiting the spectrum of possible honeywords.

Yu et al. [36] proposed to generate honeywords using a password-guessing model [34], which is based on an enhanced Generative Adversarial Network. They evaluated their HGT quantitatively and qualitatively, demonstrating that their HGT could generate honeywords more resistant to trawling attacks than other state-of-the-art HGTs.

For targeted honeyword generation, the challenge is to split the real password into tokens while retaining tokens that correspond to PII and replacing tokens that do not correspond to PII with random ones. Consider the real password *'Elena1986@327"*, the challenge is to produce honeywords containing the token *"Elena"*, which is the user's first name as indicated by her email address. To do this, we propose to employ a chunking algorithm [32] to divide passwords into semantic chunks consisting of frequently occurring sequences of related characters, and a pre-trained generative model [4] to create desired honeywords based on the semantic chunks retrieved from the chunking step.

1.3   Our Contribution

- We are the first to use generative language models to create honeywords that are robust to targeted attacks. We propose a novel HGT, termed *Chunk-GPT3*[1] which generates honeywords by segmenting passwords into semantic chunks and then instructing GPT-3 to construct honeywords containing the given semantic chunks. Without being trained on real passwords, the off-the-shelf GPT-3 model could generate high-quality honeywords that are more indistinguishable from literature counterparts, and thus are more robust to targeted attacks. Furthermore, unlike HGTs from the literature, our model makes no assumptions as to the PII an attacker may use to tell apart honeywords from the real password.
- We are the first to take semantic meaning into consideration to evaluate HGTs. We propose *HWSimilarity*, for measuring an HGT's capabilities. HWSimilarity employs a pre-trained language model MPNet [23] to encode sweetwords into vectors, and then calculates the cosine similarity between each honeyword vector and its real password vector, taking into consideration the semantics of each sweetword.
- We evaluated the capabilities of Chunk-GPT3 and two state-of-the-art HGTs and demonstrated that Chunk-GPT3-generated honeywords are significantly more similar to their real passwords, making them more difficult to differentiate regardless of what PII is available in a targeted attack.

The remainder of the paper is structured as follows: Sect. 2 provides the preliminaries for understanding our work. Section 3 introduces our approach to generating honeywords in a targeted manner. Section 4 evaluates our HGT and other two approaches. Section 5 discusses the limitations of our work and future directions. Section 6 concludes our work.

---

[1] Source code: https://github.com/HumanMachineLab/Chunk-GPT3.

2   Preliminaries

In this section, we explain the honeyword generation mechanism and datasets used in this paper.

2.1   The Honeyword Mechanism

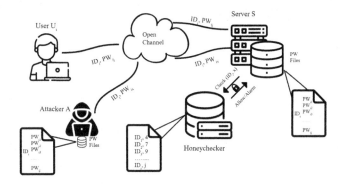

Fig. 1. Password (PW) authentication with honeywords.

Juels and Rivest [13] are the first to introduce the honeyword concept to detect password breaches. The honeyword system is comprised of four entities, as shown in Fig 1 [29]: a user U_i, an authentication server S, a *honeychecker*, and an attacker A. User U_i initially registers an account(ID_i, PW_i) on the server S. Apart from the standard user registration processes, S runs a command $GEN(k, PW_i)$ to produce a list of $k - 1$ unique fake passwords (called honeywords) to be stored alongside U_i's true password PW_i, where $k = 20$ as recommended in [13]. PW_i and its $k - 1$ honeywords are referred to as k sweetwords.

2.2   Threat Model

Honeyword-enabled systems could reliably identify a password file leak by pairing each user's account with $k - 1$ honeywords. The reason for this is that even if attackers obtain a copy of the password file along with its hashing parameters and salts, and successfully recover all the passwords via brute-force or other password guessing techniques [17,30] (be aware that at this stage they know which k sweetwords are associated with each user), they must first distinguish each user's true password from these k sweetwords. The system features *honeychecker* to aid in the usage of honeywords, and the computer system could interact with the *honeychecker* whenever a login attempt is made or users change their passwords. Additionally, the *honeychecker* is capable of triggering an alert if an anomaly is discovered. The warning signal may be sent to an administrator or to a third party [13]. This approach is compatible with existing authentication systems since it needs little adjustments to the server-side systems and no alterations to the client-side systems; nevertheless, it is very reliable due to the high probability of capturing adversaries. For instance, if the likelihood of an attacker selecting each sweetword is uniform, the probability of capturing an attacker is $3/4 = 75\%$ for $k = 4$, and thus the probability grows as k increases.

Our HGT is designed based on the assumption that attackers have complete knowledge of users' PII, and our technique including the specifics (the following mentioned prompt and temperature). As described in Sect. 5, we ensure that our honeyword generation process is irreversible even when attackers have all of the aforementioned information.

2.3   Dataset

This section introduces the password dataset (termed 4iQ) we used in this paper and password selection process. 4iQ contains a leaked compilation of various password breaches over time and was first discovered in the Dark Web[2] in December 2017. The dataset consists of 1.4 billion email-password pairs, with 1.1 billion unique emails and 463 million unique passwords. Duplicate email-password pairs were removed by an unknown curator. The listed leaks are from websites such as Canva, Chegg, Dropbox, LinkedIn, Yahoo!, etc. We eliminated the suffix of each email address and only use the prefix as usernames for simplification.

To acquire legitimate passwords, we excluded those that are too short or too lengthy, with fewer than 8 characters or more than 32 characters, respectively [27], resulting in 28,492 username-password pairs. Such short strings are not permitted by most authentication systems [24], and such lengthy strings are unlikely created by users or password managers owing to their default settings of 12, 16 or 20 characters (LastPass, 1Password and Dashlane) [32]. We further calculated the strength of each password using zxcvbn [31], and found that 24,661 passwords have a zxcvbn score of 4, 2706 passwords have a zxcvbn score of 3, 277 and 3 passwords have a zxcvbn score of 1 and 0, respectively.

To compare HGTs' capability on various password strengths, we constructed two sets of username-password combinations depending on the computed zxcvbn password strength. One *zxcvbn-weak* set with 1000 username-password pairs whose passwords have the lowest zxcvbn score, and one *zxcvbn-strong* set with 1000 username-password pairings whose passwords have the highest strength zxcvbn score. Note that all passwords in the zxcvbn-strong set have a zxcvbn score of 4, and the zxcvbn-weak set has passwords with score ranging from 0 to 2. We further analyzed and compared the chunks in the two sets and generated honeywords for both sets with our proposed method and two other HGTs.

---

[2] 1.4 Billion Clear Text Credentials Discovered in a Single Database: https://mediu m.com/4iqdelvedeep/1-4-billion-clear-text-credentials-discovered-in-a-single-datab ase-3131d0a1ae14.

3   Our Methodology

To preserve the PII in honeywords, it is necessary to segment passwords into chunks in which the PII is included. The chunks can then be used as inputs for a generative language model to produce honeywords that retain the PII while altering the real passwords.

3.1   PII Extraction

PII is rarely a single character. Instead, most PII, such as usernames, birthdays, anniversaries, and pet names carry some semantics. Semantic chunks in passwords may or may not constitute PII, but if users construct passwords including semantic chunks, they risk exposing PII. In order to extract PII from users' real passwords, we first segment the real passwords into semantic chunks using the password-specific segmentation technique PwdSegment [32]. PwdSegment conceptually trains a Byte-Pair-Encoding (BPE) for producing chunk vocabularies using training data of plain-text passwords. The BPE algorithm, which was initially proposed in 1994 as a data compression technique, is widely used in the NLP domain for subword segmentation (e.g., the GPT-2 model [18] proposed by OpenAI and the RoBERTa model [16] proposed by Meta), which preserves the frequent words while dividing the rare ones into multiple units. PwdSegment enhances the BPE technique by substituting the number of merging operations with the configurable parameter average length (avg_len) of chunk vocabulary. PwdSegment counts all character pairs and terminates the merging operation when the avg_len of the resultant chunk vocabulary equals or exceeds the threshold length. PwdSegment could be parameterized with a threshold avg_len to control the segmentation result with varied granularity more simply where a longer avg_len yields a more coarse-grained result.

The PwdSegment algorithm is first trained using a plain-text corpus. Then it repeatedly merges the most common pair of tokens into a single, new (i.e., previously unseen) token comprising the subword (i.e. chunk) vocabulary. Every merging procedure generates a new chunk by exchanging the most common pair of letters or character sequences (for example, "r", "d") with a new subword (for example, "rd"). The merging procedure is repeated until avg_len of the resultant chunk vocabulary equals or exceeds a pre-determined threshold length.

3.2   Chunk Analysis for zxcvbn-weak and zxcvbn-strong Password Sets

Difference of Chunk Numbers. We segment passwords into chunks for both zxcvbn-weak and zxcvbn-strong password sets using the PwdSegment algorithm. As shown in Fig 2, most passwords in the zxcvbn-strong set contain four to seven chunks, whereas most passwords in the zxcvbn-weak passwords only contain two or three chunks. This suggests that stronger passwords (based on zxcvbn) typically contain more chunks than weak passwords.

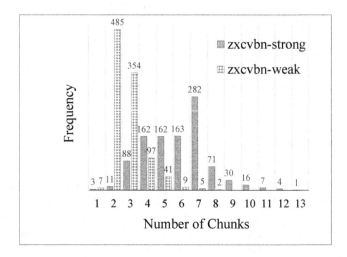

Fig. 2. The comparison of password chunk numbers in zxcvbn-weak and zxcvbn-strong sets.

Difference of Common Chunk Frequencies. To further investigate the differences between zxcvbn-strong password set and zxcvbn-weak password set, we list all chunks in both sets and visualize the result in Fig 3, from which we can observe that most chunks in the zxcvbn-weak password set contain semantics or easy-to-guess patterns, such as English words ("football", "builder" "vietnamese", "microsoft"), phrases ("iloveyou"), Chinese names ("chenchen", "liang", "jiang", "shan"), English names ("benjamin", "Erick", "sasha", "elena"), and patterns ("qwert", "zxcvbn", "QWEASDZXC"). Many of them are plausible PII that attackers could take advantage of to compromise users' accounts. In contrast, the majority of chunks in the zxcvbn-strong password set are random and short combinations of characters with no semantics, whereas semantics still exist in certain chunks (such as "sasha", "jj" and "wang"). This indicates that although passwords that are zxcvbn-strong in strength are mostly comprised of more chunks and are harder to guess in a trawling scenario, many of them still contain semantic words, which can be PII that is accessible to attackers, thereby increasing the likelihood of passwords being guessed and accounts being compromised. As a result, regardless of the strength of the real password, as long as it contains PII which attackers could utilize all their resources to get, the trawling-honeyword-integrated authentication system will fail since most trawling-generated honeywords do not contain PII, and thus a targted-honeyword-integrated system is needed.

3.3 Honeyword Generation with Chunk-GPT3

Language models can learn the probabilities of occurrences of a series of words in a regularly spoken language and predict the next potential word in that sequence. Generative Pre-trained Transformer 3 (GPT-3) is an autoregressive language

Fig. 3. The comparison of common password chunks in zxcvbn-weak (left) and zxcvbn-strong (right) sets.

model that uses deep learning to generate text that appears to be written by a person. It was introduced in 2020 and excels at a variety of NLP tasks, including translation, question-answering, and cloze [4]. The model was trained on trillions of words in text documents. It turns words into vectors or mathematical representations, and then decodes the encoded text into human-readable phrases. The model can be utilized to execute NLP tasks without requiring fine-tuning on particular downstream task datasets and is capable of producing texts that are difficult for humans to differentiate from human-written articles.

Therefore, we propose to use GPT-3 to generate honeywords that are robust to targeted attacks by providing the semantic chunks retrieved in the PII extraction phase. We first specify what the model should do by giving it a prompt, for example, "Derive five passwords that are similar to '*toby*2009*bjs*' and contain '*toby*', '2009' and '*bjs*'. Do not add digits at the end of the passwords." Here, "*toby*", "2009" and "*bjs*" are chunks generated by PwdSegment. GPT-3 will then produce outputs "*tobyEmma2009bjs, toby2009Katiebjs, toby2009bjsKaitlyn, toby2009bjsRiley, toby2009bjsSavannah*" by following the instruction. The quality and the diversity of the output depend on three attributes: prompt, temperature and examples given to the model.

The Prompt. The prompt is the instruction GPT-3 received. The quality of the prompt can determine the quality of the generated honeywords. Usually, the more concise and instructive the prompt is, the better the completion is [15]. Same can be seen in honeyword generation, as shown in Table 2.

Table 2. Honeywords generated by GPT-3 when using different prompts. Honeywords generated using Prompt2 are not ideal because they do not contain the potential PII "toby".

Prompt1	Suggest three passwords that are similar to "toby2009bjs" and contain "toby"
Honeywords	toby2009bjd, toby1998bjx, toby2021bjz
Prompt2	Suggest three passwords that look like "toby2009bjs".
Honeywords	toy2009bjs, tab2009bjs, boy2009bjs

The Temperature. The temperature is a numeric variable between 0 and 1 that effectively regulates the model's degree of confidence when generating predictions. A lower temperature implies that the model will take fewer risks, and the honeywords created will be more repetitive while increasing the temperature results in more diversified honeywords. The temperature is a vital parameter that determines the irreversibility of our HGT, as discussed in Sect. 5. Table 3 contains examples of honeywords formed at temperatures 0 and 1.

Table 3. Honeywords generated by GPT-3 when using different temperatures and given the prompt "Suggest five words that are similar to 'toby2009bjs' and contain 'toby'." A higher temperature will result in more diverse honeywords.

Temperature	Honeywords
0	toby2009bjd, toby2009bjx, toby2009bjz, toby2009bjf, toby2009bjh
1	Toby2009BJS, toby2009bjs1, tobybjs2009, Bjs2009toby, bjs2009toby1

Zero-Shot and Few-Shot Learning. Zero-shot learning refers to a situation in which no demonstrations are permitted and the model is given simply a plain language description of the task. In comparison, few-shot learning refers to a situation in which the model is given a few demonstrations of the task during inference time, but the model is not re-trained on them. This is particularly advantageous since many websites have varying policies regarding password creation, such as beginning with letters and requiring uppercase, lowercase, symbols, and numbers. When the operators demonstrate how they want the honeywords to appear, GPT-3 will generate honeywords that match the examples.

Since the introduction of Generative Pre-trained Transformers, they have been extensively investigated in a variety of domains, including creating media dialogues summaries [7], generating code from natural-language instructions [6], generating passphrases [11], and generating graphics from text descriptions [19]. To the best of our knowledge, we are the first to employ GPT-3 in the sphere of computer security, to generate honeywords that are resistant to targeted attacks.

An example of honeyword generation using Chunk-GPT3 is illustrated in Fig. 4, which contains two steps: 1). Passwords are segmented using algorithm PwdSegment, detailed in Sect. 3.1. For example, password "Elena1986@327" is segmented into chunks "Elena", "1986" and "327". 2). The resulting chunks are used as inputs to prompt GPT-3 to generate honeywords. We prompted GPT-3 with instruction "Please derive three passwords that are similar to "Elena1986@327" and contain "Elena", "1986" and "327". The length of the passwords should be at most 13 (the length of the real password "Elena1986@327")."

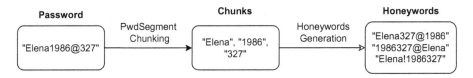

Fig. 4. Honeyword generation with Chunk-GPT3. In this example, the password "Elena1986@327" is segmented into chunks "Elena", "1986", and "327" using the PwdSegment Chunking algorithm. The chunks are then used as inputs for GPT-3 to generate honeywords.

4 Evaluation

Two common metrics in HGT evaluation are *flatness* and *success-number graphs* which measure HGTs' resistance against the honeyword distinguishing attacker from the average and worst-case point perspective [26]. The honeyword distinguishing attacker is required for using the two metrics. Previous works [9,10] used the trawling attack algorithm Normalized Top-PW model to construct *flatness* and *success-number graphs* and to evaluate their HGTs, since their HGTs are used to generate honeywords against trawling attacks [35]. The Normalized Top-PW is not applicable to targeted attacks because trawling attackers have no knowledge about users' PII while targeted attacks do, which make targeted attackers more capable. To the best of our knowledge, the only work proposing targeted attacks [29] construct their attack models based on various kinds of capabilities allowed to an attacker (e.g., birthday, username, email address, and registration order). We do not give these assumptions to attackers since it is typically not know what kind of attackers a system may have when generating honeywords. In fact, attackers may take advantage of any resources they may have, not limiting to PII, registration order and more. A comparison of the assumptions made in our HGT and in Wang et al.'s is shown in Table 4. Wang et al. [29] used flatness and success-number graphs to measure their HGTs. These metrics measure password guessing success rate per user, and the number of successfully identified real passwords, respectively. However, these honeyword evaluation metrics are not compatible with our method since flatness and success-number graphs require the computation of password probabilities as yielded by the HGT method. For example, if our honeywords were generated

using a PCFG-based approach, we would be able to compute honeyword probabilities. However, Chunk-GPT3 is not a probabilistic method but a generalized
application of a large language model over password chunks. Thus, we evaluate
honeywords from the perspective of word embedding similarities, which are commonly used in the NLP domain, to measure the similarity of two sequences. We
propose an evaluation metric that measures the effectiveness of HGTs by comparing the similarity between a honeyword and its real password using another
pre-trained language model. We also intend to draw the community's attention
to targeted scenarios, since trawling situations have been intensively studied
but targeted honeyword generation and attack models are under-researched yet
represent a pressing problem, as outlined in Sect. 1.1 and in [28].

Table 4. Assumptions on attackers in our HGT and Wang et al.'s [29].

	PW file	Public info[a]	Limited PII or user info	Any PII or other info
Ours	✓	✓	✓	✓
Wang et al.'s [29]	✓	✓	✓	

[a] Public info may include leaked password lists, password policy, and cryptographic
algorithms.

4.1   Metric: HWSimilarity

In this section, we introduce an evaluation metric to measure the indistinguishability of honeywords in terms of their corresponding real passwords.

The similarity between two strings is crucial in HGT since it demonstrates the
indistinguishability of a false password from a genuine one. Typically, in natural
language processing tasks, the distance/similarity of two strings is determined as
follows: the strings are converted to vectors using word embedding techniques,
and then the cosine similarity of the two vectors is calculated as the distance.
Here, the strings might be composed of letters, symbols, or numbers, similar to
how passwords are composed. Since passwords may contain PII which contains
semantics, hence when measuring the similarity of two sweetwords, the semantics
contained in a sweetword have to be considered. Therefore, in this paper, we propose to use a pre-trained language model MPNet [23] to encode passwords since
it encodes the semantics in word sequences to word embeddings. MPNet utilizes
the interdependence among predicted tokens via permuted language modeling
(vs. MLM in BERT [8]) and accepts auxiliary position information as input to
help the model view a whole phrase, hence minimizing position discrepancy (vs.
PLM in XLNet [33]).

Computing the HWSimilarity of a sweetword list can be done as follows:
For a user's sweetword list $SW = [sw_1, sw_2.....sw_l]$, and her honeyword list
$HW = [hw_1, hw_2.....hw_{l-1}]$, we have $pw \in SW, HW \subsetneq SW$ and $pw \notin HW$.

Here pw is the user's real password, hw_i denotes a honeyword and l is the number of sweetwords. $HW\,Similarity = \frac{\sum_{i=1}^{l-1} cosin(\Phi(hw_i),\Phi(pw))}{l-1}$, here Φ is the MPNet Neural Network model.

4.2   Comparable HGTs and Evaluation Results

We compare our Chunk-GPT3 with other three HGTs: generating honeywords using GPT-3 alone without semantic chunks provided, and two state-of-the-art HGTs chaffing-by-tweaking and *fasttext*.

Chaffing-by-Tweaking. Chaffing-by-tweaking (tweaking) HGT was initially presented in [13] and mainly relies on random letter, digit, and symbol substitution. We choose to use chaffing-by-tweaking instead of other recently proposed methods in the literature because other methods are more vulnerable to targeted attacks [2,28]. Dionysiou et al. [9] highlight the intricacy of developing tweaking rules in such a way that it could be difficult for an attacker to distinguish the password from its changed versions. For example, if a chaffing-by-tweaking strategy randomly perturbs the last three characters of a password, the adversary may easily conclude that the authentic password is the first one in the instances "18!morning", "18!morniey", and "18!gorndge". Thus, they replace all occurrences of a particular symbol in a given password with a randomly chosen alternate symbol, lower-case each letter in a password with probability $p = 0.3$, upper-case each letter in a password with probability $f = 0.03$, and replace each digit occurrence with probability $q = 0.05$. [9] contains the pseudocode and rationale for the assignment of $p, q,$ and f.

Table 5. Honeyword samples generated by the HGTs compared in the paper (Chunk-GPT3, GPT-3, *fasttext* and tweaking). *fasttext* is required to be trained on a real password dataset (the *rockyou* dataset in the paper). Other three HGTs can generate honeywords directly without being trained on a password dataset. Only Chunk-GPT3-generated honeywords retain the PII in the real password.

	HGTs			
	Chunk-GPT3	**GPT-3**	*fasttext*	**tweaking**
h2omega-tania	tania-home123	h2omega-alex	Karert_334	4oMega<tANia
	Tania@home5	h2omega-zoe	Adery993	H2oMega"tAnia
	home!tania12	h2omega-sam	brobe31	h4omega,tania
0000_mila_0000	1111_mila_0000	0000_lila_0000	octavia3	7434~MIla$6421
	0000_MILA_0000	0000_lela_0000	Bushido07	364\MIlA-9353
	0000@Mila@0000	0000_lola_0000	Dampire2	3124/MiLa'2089
007skyblueboy	Skyblueboys007	007skybluegirl	gz152sha	903SkyBlUeboY
	Blueboysky007	007babyblueboy	Calepepi	561SkYblUEbOy
	007blueboysky	007lightblueboy	hajenrai	960SKybluebOy

Chaffing-by-Fasttext. This technique was proposed by Dionysiou et al. [9] which uses representation learning for the generation of honeywords. They convert words to vectors using *fasttext* and then assign honeywords to the $k - 1$ nearest neighbors of an actual password based on cosine similarity.

More specifically, in the chaffing-by-fasttext method, it needs a real password corpus as the training dataset for the *fasttext* model. During the training phase, *fasttext* generates vector representations of each word in the corpus. After training is complete, the trained model can be queried by providing a real password as input and receiving a multi-dimensional vector representing the provided password's word embedding as a response. Following that, Dionysiou et al. loop over each password in their password corpus (n records in total where n is the number of users) and return its top $k - 1$ closest neighbours in decreasing order of cosine similarity to create the list of $k \times n$ sweetwords. In this way, for each password in the password file, they generate a list of the $k - 1$ most similar honeywords.

Notably, the technique's primary weakness is that the produced honeywords are all genuine passwords in the *fasttext* training dataset, which means that if an attacker has access to the training dataset, the honeywords will be readily discovered. Additionally, the size of the training data has a significant impact on the quality of the honeywords created.

GPT-3 Without Semantic Chunks. We conducted an ablation study to assess if GPT-3 can create honeywords containing PII on its own, without any semantic chunks provided. In this case, the prompt we gave GPT-3 is "Derive 19 passwords that are similar to *real_password*. The length of the passwords should be at most *len(real_password)*. Do not add digits at the end of the passwords."

A few examples of honeywords generated by Chunk-GPT3, GPT-3, tweaking and *fasttext* are illustrated in Table 5.

Table 6. HWSimilarity of honeywords generated by the four techniques (Chunk-GPT3, GPT-3, *fasttext*, and tweaking). Honeywords generated by Chunk-GPT3 have the highest HWSimilarity score compared with other HGTs, indicating that the Chunk-GPT3-generated honeywords are the most similar to their cor- responding real passwords taking into account semantics.

	Chunk-GPT3	GPT-3	*fasttext*	tweaking
zxcvbn-strong	0.8525	0.8348	0.3441	0.7297
zxcvbn-weak	0.8367	0.8144	0.3445	0.7527

Results. The HWSimilarity of honeywords is shown in Table 6. For both zxcvbn-strong and zxcvbn-weak password sets, honeywords generated by *fasttext* and tweaking have a much lower HWSimilarity score than the score of honeywords generated by GPT-3 and Chunk-GPT3, indicating that the majority of *fasttext* and tweaking-generated honeywords do not contain users' PII.

We also compared GPT-3 and Chunk-GPT3 using paired t-tests, and found the Chunk-GPT3-generated honeywords are significantly more similar to their corresponding real passwords considering semantics contained in passwords, and thus are harder to distinguish by targeted attacks ($t_{zxcvbn-weak}(999) = 3.935, P < 0.001$, $t_{zxcvbn-strong}(999) = 3.237, P < 0.001$).

Will HWSimilarity Leak Information About the Real Passwords to Attackers? Consider this scenario: An attacker takes the 20 sweetwords and creates 20 different sets S_1, S_2, ..., and S_{20} of 19 sweetwords each (i.e., leaving a different sweetword out every time). Then for each of S_1, S_2, ..., and S_{20}, the attacker computes the HWSimilarity of each element of S_i against the sweetwords that are not in S_i. Will this expose some patterns revealing which of the 20 sweetwords is the real password? In order to examine this, we did a pilot experiment and took a subset of our data with 500 username-password pairs and 4 honeywords per user from the generated honeywords by the 4 HGTs. In this case, in the sweetword files with honeywords generated by different HGTs, each user has 4 honeywords stored along with the real passwords. For each sweetword list, the attacker takes one sweetword as p, and then 1) Computes the average HWSimilarity score ($\bar{p}$) of each sweetword sw_1 to sw_4 against the target sweetword p. 2) Then computes the average HWSimilarity score ($\bar{a}_1$) of sweetwords sw_2, sw_3, sw_4, and p against sw_1. 3) Next, computes the average HWSimilarity score ($\bar{a}_2$) of sweetwords sw_1, sw_3, sw_4, and p against sw_2. 4) Then computes the average HWSimilarity score ($\bar{a}_3$) of sweetwords sw_1, sw_2, sw_4, and p against sw_3. 5) Then computes the average HWSimilarity score ($\bar{a}_4$) of sweetwords sw_1, sw_2, sw_3, and p against sw_4. 6) Finally, checks if one of the values (i.e., $\bar{p}, \bar{a}_1, \bar{a}_2, \bar{a}_3, \bar{a}_4$) is significantly "different" from the other 4 values.

The average similarity scores for each HGT are shown in Fig 5. ANOVA tests on each HGT's averages did not reject the null hypotheses, concluding that there is no significant difference between the averages, suggesting that HWSimilarity would not reveal the real password.

5 Discussion

We talk about the limitations of our study and future directions in this section.

User Study. We argue that there is no need to conduct user studies to qualitatively evaluate Chunk-GPT3-generated honeywords. If given a question: "Suppose you are an attacker and know a victim's user name is 'mila', which one in the following list would most probably be his/her password: '$0000\_mila\_0000, octavia3, Bushido07, Dampire2$' (real password and honeywords generated by *fasttext*).", the task is easy to complete, while if the choices are "$0000\_mila\_0000, 1111\_mila\_0000, 0000\_MILA\_0000, 0000@Mila@0000$" (real password and honeywords generated by Chunk-GPT3), the task becomes obviously more difficult.

Lack of Comparison with [29]. To the best of our knowledge, there is only one publication that discusses how to generate honeywords that are resistant

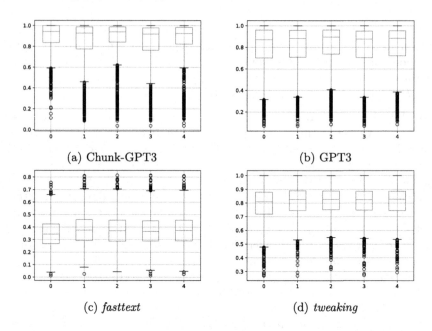

(a) Chunk-GPT3   (b) GPT3

(c) fasttext   (d) tweaking

Fig. 5. Each boxplot represents the HWSimilarity scores of sweetwords at all indices with the sweetword at the target index. No significant difference in the average scores at different indices is observed for each HGT.

to targeted attacks, which was published in IEEE S&P'22 by Wang et al. [29]. They first proposed four attack models each representing a potential attacker A's strategy, with each model based on different information available to A (e.g., public datasets, the victim's username, email address, birthday and registration order). They further developed four HGTs for each attack strategy, by using various probabilistic password guessing models proposed in previous work [28]. Nonetheless, assuming that attackers have access to only certain PII imposes an important limit since attackers may utilize a superset of PII beyond the PII pieces considered in their study (or a totally different PII set) to guess a user's password, particularly if the user is a person of interest. What we are proposing is a different yet robust, and generalized approach. Rather than assuming A's attack strategy and creating HGTs accordingly, we assume attackers have white-box access to our HGT, meaning that attackers have complete knowledge of users' PII, and our technique including the specifics (such as the prompt and temperature). A comparison of the assumptions made in our HGT and in Wang et al.'s can be found in Table 4, along with an explanation as to why generating flatness and success-number graphs is not possible (see Sect. 4). Despite the impossibility of producing a useful comparison between our work and Wang et al.'s, and for completeness, we still attempted to reproduce their HGT and compare with ours, using HWSimilarity. However, they did not make their artifacts public due to intellectual property concerns, and despite our efforts, we were unable to reproduce their HGT from the description found in their paper. Nonetheless, it is clear that their HGT method does not consider the real password chunks, and this results in honeywords that do not necessarily resemble the real password. This can be readily seen by comparing the honeywords generated with one of their HGTs (TarList) and our Chunk-GPT3 method (see Table 7).

Table 7. Honeyword examples generated for real password "tiger81" by our method (Chunk-GPT3) and Wang et al.'s TarList (taken from [29], Fig. 1).

HGTs	Honeywords
Chunk-GPT3	Tig3r81, T1ger81, TigEr81, Tig3r8I, T1g3r81, Tig3r1I, T1gEr81, T1ger8I, TigEr1I, Tig3r8I, T1gEr8I, T1g3r1I
Wang et al.'s	jsmith117, prince00, love123, qwertyu, js128821, bond007, a123456, trustono1, rcv_11n1nj, jan1981, lemein, newy0rk, 1989y2002r

Since our HGT is based on the intuition that honeywords that are more similar to their corresponding real password are of higher quality [13], if there are any PII in the real passwords, the honeywords should include that PII to warrant their indistinguishability. Thus, we evaluate our HGT based on the similarity/word vector distance between honeywords and real passwords.

Irreversibility. The irreversibility of an HGT is critical. We need to make sure that even when attackers know our methodology and the specifications we were using for generating honeywords, such as the prompt and the temperature, they still cannot reproduce the honeywords we generated. This is ensured by careful prompt-engineering [14,21] and temperature setting. We suggest to set temperature to 1 to get the most randomness [4], and after experimenting with various prompts, we decided to use the prompt "Derive 19 passwords that are similar to *real_password*, and contain *chunks*. The length of the passwords should be at most $len(real\_password)$. Do not add digits at the end of the passwords." since it generates the most diversified honeywords compared with other prompts we experimented with, and the honeywords generated each time are different by our observation.

6   Conclusions

In this paper, we proposed a novel HGT, Chunk-GPT3, which segments passwords into semantic chunks and then utilizes GPT-3 to generate high-quality honeywords that contain PII existing in users' real passwords. Honeywords generated by Chunk-GPT3 are robust to targeted attacks where attackers get access to both breached password databases and users' PII. Unlike other machine learning-based HGTs, GPT-3 can be easily integrated into any current password-based authentication system without any further training on real passwords. Additionally, we proposed a targeted HGT evaluation metric that incorporates another pre-trained language model. We compared Chunk-GPT3's performance with GPT-3 alone, and two state-of-the-art HGTs with the proposed metric and demonstrated that Chunk-GPT3-generated honeywords are significantly harder to decipher and thus could raise the bar for targeted attacks.

Acknowledgement. The authors thank the assigned shepherd and anonymous reviewers for their valuable comments that improved the quality of the paper. We acknowledge the support of the Natural Sciences and Engineering Research Council of Canada (NSERC), funding reference number RGPIN-2018-05919.

References

1. IBM security: Cost of a data breach report 2021 (2021). https://www.ibm.com/security/data-breach. Accessed 01 Jan 2022
2. Bojinov, H., Bursztein, E., Boyen, X., Boneh, D.: Kamouflage: loss-resistant password management. In: Gritzalis, D., Preneel, B., Theoharidou, M. (eds.) ESORICS 2010. LNCS, vol. 6345, pp. 286–302. Springer, Heidelberg (2010). https://doi.org/10.1007/978-3-642-15497-3_18
3. Bonneau, J., Herley, C., Oorschot, P.C.V., Stajano, F.: The quest to replace passwords: a framework for comparative evaluation of web authentication schemes. In: 2012 IEEE Symposium on Security and Privacy (S&P), pp. 553–567 (2012). https://doi.org/10.1109/SP.2012.44
4. Brown, T., et al.: Language models are few-shot learners. In: Advances in Neural Information Processing Systems, vol. 33, pp. 1877–1901. Curran Associates, Inc. (2020). https://proceedings.neurips.cc/paper_files/paper/2020/file/1457c0d6bfcb4967418bfb8ac142f64a-Paper.pdf
5. Camenisch, J., Lehmann, A., Neven, G.: Optimal distributed password verification. In: Proceedings of the 22nd ACM SIGSAC Conference on Computer and Communications Security. CCS '15, pp. 182–194. Association for Computing Machinery, New York, NY, USA (2015). https://doi.org/10.1145/2810103.2813722
6. Chen, M., et al.: Evaluating large language models trained on code. arXiv preprint arXiv:2107.03374 (2021)
7. Chintagunta, B., Katariya, N., Amatriain, X., Kannan, A.: Medically aware GPT-3 as a data generator for medical dialogue summarization. In: Proceedings of the Second Workshop on Natural Language Processing for Medical Conversations, pp. 66–76. Association for Computational Linguistics, Online, June 2021). https://doi.org/10.18653/v1/2021.nlpmc-1.9
8. Devlin, J., Chang, M.W., Lee, K., Toutanova, K.: BERT: pre-training of deep bidirectional transformers for language understanding. In: Proceedings of the 2019 Conference of the North American Chapter of the Association for Computational Linguistics: Human Language Technologies, Volume 1 (Long and Short Papers), pp. 4171–4186. Association for Computational Linguistics, Minneapolis, Minnesota, June 2019. https://doi.org/10.18653/v1/N19-1423
9. Dionysiou, A., Vassiliades, V., Athanasopoulos, E.: HoneyGen: generating honeywords using representation learning. In: Proceedings of the 2021 ACM Asia Conference on Computer and Communications Security. ASIA CCS '21, pp. 265–279. Association for Computing Machinery, New York, NY, USA (2021). https://doi.org/10.1145/3433210.3453092

10. Guo, Y., Zhang, Z., Guo, Y.: Superword: a honeyword system for achieving higher security goals. Comput. Secur. **103**, 101689 (2021). https://doi.org/10.1016/j.cose.2019.101689

11. Jagadeesh, N., Vargas Martin, M.: Alice in passphraseland: assessing the memorability of familiar vocabularies for system-assigned passphrases (2021). https://doi.org/10.48550/ARXIV.2112.03359

12. Joudaki, Z., Thorpe, J., Vargas Martin, M.: Reinforcing system-assigned passphrases through implicit learning. In: Proceedings of the 2018 ACM SIGSAC Conference on Computer and Communications Security. CCS '18, pp. 1533–1548. Association for Computing Machinery, New York, NY, USA (2018). https://doi.org/10.1145/3243734.3243764

13. Juels, A., Rivest, R.L.: Honeywords: making password-cracking detectable. In: Proceedings of the 2013 ACM SIGSAC Conference on Computer and Communications Security. CCS '13, pp. 145–160. Association for Computing Machinery, New York, NY, USA (2013). https://doi.org/10.1145/2508859.2516671

14. Kojima, T., Gu, S.S., Reid, M., Matsuo, Y., Iwasawa, Y.: Large language models are zero-shot reasoners. arXiv preprint arXiv:2205.11916 (2022)

15. Liu, P., Yuan, W., Fu, J., Jiang, Z., Hayashi, H., Neubig, G.: Pre-train, prompt, and predict: a systematic survey of prompting methods in natural language processing. ACM Comput. Surv. **55**(9) (2023). https://doi.org/10.1145/3560815

16. Liu, Y., et al.: RoBERTa: a robustly optimized BERT pretraining approach. arXiv preprint arXiv:1907.11692 (2019)

17. Pasquini, D., Gangwal, A., Ateniese, G., Bernaschi, M., Conti, M.: Improving password guessing via representation learning. In: 2021 IEEE Symposium on Security and Privacy (S&P, pp. 1382–1399 (2021). https://doi.org/10.1109/SP40001.2021.00016

18. Radford, A., Wu, J., Child, R., Luan, D., Amodei, D., Sutskever, I.: Language models are unsupervised multitask learners (2018). https://d4mucfpksywv.cloudfront.net/better-language-models/language-models.pdf

19. Ramesh, A., et al.: Zero-shot text-to-image generation. In: International Conference on Machine Learning, pp. 8821–8831. PMLR (2021)

20. Ratha, N.K., Connell, J.H., Bolle, R.M.: Enhancing security and privacy in biometrics-based authentication systems. IBM Syst. J. **40**(3), 614–634 (2001)

21. Reynolds, L., McDonell, K.: Prompt programming for large language models: beyond the few-shot paradigm. In: Extended Abstracts of the 2021 CHI Conference on Human Factors in Computing Systems. CHI EA '21, Association for Computing Machinery, New York, NY, USA (2021). https://doi.org/10.1145/3411763.3451760

22. Roche, T., Lomné, V., Mutschler, C., Imbert, L.: A side journey to Titan. In: 30th USENIX Security Symposium (USENIX Security 21), pp. 231–248. USENIX Association, August 2021. https://www.usenix.org/conference/usenixsecurity21/presentation/roche

23. Song, K., Tan, X., Qin, T., Lu, J., Liu, T.Y.: MPNet: masked and permuted pre-training for language understanding. In: Advances in Neural Information Processing Systems, vol. 33, pp. 16857–16867. Curran Associates, Inc. (2020). https://proceedings.neurips.cc/paper_files/paper/2020/file/c3a690be93aa602ee2dc0ccab5b7b67e-Paper.pdf

24. Tan, J., Bauer, L., Christin, N., Cranor, L.F.: Practical recommendations for stronger, more usable passwords combining minimum-strength, minimum-length, and blocklist requirements. CCS '20, pp. 1407–1426. Association for Computing Machinery, New York, NY, USA (2020). https://doi.org/10.1145/3372297.3417882

25. Thomas, K., et al.: Data breaches, phishing, or malware? Understanding the risks of stolen credentials. In: Proceedings of the 2017 ACM SIGSAC Conference on Computer and Communications Security. CCS '17, pp. 1421–1434. Association for Computing Machinery, New York, NY, USA (2017). https://doi.org/10.1145/3133956.3134067

26. Wang, D., Cheng, H., Wang, P., Yan, J., Huang, X.: A security analysis of honeywords. In: Network and Distributed System Security (NDSS) Symposium 2018, pp. 1–16, October 2018. https://doi.org/10.14722/ndss.2018.12345

27. Wang, D., Wang, P., He, D., Tian, Y.: Birthday, name and bifacial-security: understanding passwords of Chinese web users. In: 28th USENIX Security Symposium (USENIX Security 19), pp. 1537–1555. USENIX Association, Santa Clara, CA, August 2019. https://www.usenix.org/conference/usenixsecurity19/presentation/wang-ding

28. Wang, D., Zhang, Z., Wang, P., Yan, J., Huang, X.: Targeted online password guessing: an underestimated threat. In: Proceedings of the 2016 ACM SIGSAC Conference on Computer and Communications Security. CCS '16, pp. 1242–1254. Association for Computing Machinery, New York, NY, USA (2016). https://doi.org/10.1145/2976749.2978339

29. Wang, D., Zou, Y., Dong, Q., Song, Y., Huang, X.: How to attack and generate honeywords. In: 2022 IEEE Symposium on Security and Privacy (S&P), pp. 966–983 (2022). https://doi.org/10.1109/SP46214.2022.9833598

30. Weir, M., Aggarwal, S., Medeiros, B.d., Glodek, B.: Password cracking using probabilistic context-free grammars. In: 2009 30th IEEE Symposium on Security and Privacy (S&P), pp. 391–405 (2009). https://doi.org/10.1109/SP.2009.8

31. Wheeler, D.L.: zxcvbn: Low-budget password strength estimation. In: 25th USENIX Security Symposium (USENIX Security 16), pp. 157–173. USENIX Association, Austin, TX, August 2016. https://www.usenix.org/conference/usenixsecurity16/technical-sessions/presentation/wheeler

32. Xu, M., Wang, C., Yu, J., Zhang, J., Zhang, K., Han, W.: Chunk-level password guessing: towards modeling refined password composition representations. In: Proceedings of the 2021 ACM SIGSAC Conference on Computer and Communications Security. CCS '21, pp. 5–20. Association for Computing Machinery, New York, NY, USA (2021). https://doi.org/10.1145/3460120.3484743

33. Yang, Z., Dai, Z., Yang, Y., Carbonell, J., Salakhutdinov, R.R., Le, Q.V.: XLNet: generalized autoregressive pretraining for language understanding. In: Advances in Neural Information Processing Systems, vol. 32. Curran Associates, Inc. (2019). https://proceedings.neurips.cc/paper_files/paper/2019/file/dc6a7e655d7e5840e66733e9ee67cc69-Paper.pdf

34. Yu, F.: Raising the bar for password crackers: improving the quality of honeywords with deep neural networks. Master's thesis, Ontario Tech University, Oshawa, Canada (2022). https://ir.library.ontariotechu.ca/bitstream/handle/10155/1593/Yu_Fangyi.pdf?sequence=1&isAllowed=y

35. Yu, F., Vargas Martin, M.: GNPassGAN: improved generative adversarial networks for trawling offline password guessing. In: 2022 IEEE European Symposium on Security and Privacy Workshops (EuroS&PW), pp. 10–18 (2022). https://doi.org/10.1109/EuroSPW55150.2022.00009

36. Yu, F., Vargas Martin, M.: HoneyGAN: creating indistinguishable honeywords with improved generative adversarial networks. In: Lenzini, G., Meng, W. (eds.) STM 2022. LNCS, vol. 13867, pp. 189–198. Springer, Cham (2023). https://doi.org/10.1007/978-3-031-29504-1_11

Cyber Physical System Security

White-Box Concealment Attacks Against Anomaly Detectors for Cyber-Physical Systems

Alessandro Erba[1,2]( ) and Nils Ole Tippenhauer[1]

[1] CISPA Helmholtz Center for Information Security, Saarbrücken, Germany
{alessandro.erba,tippenhauer}@cispa.de
[2] Saarbrücken Graduate School of Computer Science, Saarland University, Saarbrücken, Germany

Abstract. Anomaly detection for cyber-physical systems is an effective method to detect ongoing process anomalies caused by an attacker. Recently, a number of anomaly detection techniques were proposed (e.g., ML based, invariant rule based, control theoretical). Little is known about the resilience of those anomaly detectors against attackers that conceal their attacks to evade detection. In particular, their resilience against white-box concealment attacks has so far only been investigated for the subset of neural network-based detectors. In this work, we demonstrate for the first time that white-box concealment attacks can also be applied to detectors that are not based on neural network solutions. In order to achieve this, we propose a generic white-box attack that evades anomaly detectors and can be adapted even if the target detection technique does not optimize a loss function. We design and implement a framework to perform our attacks, and test it on several detectors from related work. Our results show that it is possible to completely evade a wide range of detectors (based on diverse detection techniques) while reducing the number of samples that need to be manipulated (compared to prior black-box concealment attacks).

1 Introduction

Cyber-Physical Systems (CPS) interact with the physical environment to accomplish a task by using sensors and actuators while applying a control strategy. Examples of such systems are Industrial Control Systems (ICS), Critical Infrastructures (such as power and water systems), and Autonomous Vehicles (AV).

The security and reliability of those systems are crucial in our society. For example, the water reaches houses through water treatment and distribution systems, which are critical infrastructures, consisting of pipes, pump stations, industrial controllers, etc. Attacks targeting those infrastructures can cause disruption (e.g., no water to houses), or harm people (e.g., contaminants in water).

Recently, anomaly detection techniques for CPS gained popularity as they allow the identification of process anomalies caused by cyber-attacks while

D. Gruss et al. (Eds.): DIMVA 2023, LNCS 13959, pp. 111–131, 2023.
https://doi.org/10.1007/978-3-031-35504-2_6

remaining legacy compliant. Different techniques were proposed in the literature to detect anomalies in CPS, system identification [3,9,19,29], Kalman filtering [2], Support Vector Machines [8], Deep learning [16,22,28] and control invariants [1,13]. Little is known about the resilience of those anomaly detection techniques against targeted manipulation, especially regarding classifier evasion [4]. If an attacker evades the anomaly detection system to conceal the true state and avoid or delay detection, can cause severe hardware damage or harm human beings. Concealment attacks are a variant of evasion attacks, in which evasion by sensor manipulation will not have a direct effect on the process [11], and can be performed in white-box and black-box settings. We refer to white-box and black-box to differentiate the knowledge of the attacker. A white-box attacker has access to a copy of the anomaly detector, which can be queried to get detection scores for a sample. A black-box attacker can not access this information.

Prior work demonstrated that generic black-box concealment attacks on general anomaly detectors are possible [12], but those limitations lead to attacks that manipulate a large number of sensors, over many samples. It is unclear how optimal those attacks are—we need a baseline to compare against. White-box concealment attacks by a less constrained, more knowledgeable attacker could provide such a baseline, but those attacks were only investigated for the specific subclass of Deep Learning based anomaly detectors [11]. Thus, the threat posed by white-box concealment attacks on general anomaly detectors is unclear, and in particular, the minimal perturbation required to achieve misclassification (by strong attackers) is unknown for each detector.

In this work, we bridge this research gap by addressing three research questions: **R1** How resilient are anomaly detectors for cyber-physical systems against white-box concealment attacks? **R2** Can white-box attacks efficiently compute manipulations at runtime? **R3** How do the white-box attacks perform compared to prior work black-box attacks?

To address the aforementioned research questions we tackle two research challenges: **C1** The attacker manipulates dynamic streaming data, i.e., the attacker cannot retroactively change past values, or predict future process sensor values. **C2** General detectors are not guaranteed to optimize a differentiable loss function for detection (in contrast to Deep Learning-based detectors). We address C1 by implementing and evaluating a method that manipulates only the current sensors' observations and show that it is still possible to minimize the detection function loss. We address C2 by proposing a method to re-write non-differentiable classification functions as differentiable and hence allow concealment attacks.

List of Contributions. The main contributions of the paper are:

- Designing an effective general purpose white-box concealment framework for anomaly detection systems.
- Formulation of loss-free detectors (i.e. process invariants), as loss-based.
- Evaluation of proposed white-box attacks with real testbed data against five state-of-the-art anomaly detection systems.
- Comparison of the proposed white-box concealment with prior attacks.

Table 1. Summary of anomaly detection families proposed in prior work in the context of CPS. The table reports the approach used for detection and the detectors that we analyze in our evaluation ($\bigcirc$ = no, $\bullet$ = yes). We skip DNN as it was analyzed before.

	[19]	[3]	[9,29]	[2,8]	[16,22,28]	[1,13]
Approach type	AR	SVD	LTI	SVM	DNN	Invariants
Classification Differentiability	D	D	D	D	D	N
Prior WBC analysis	$\bigcirc$	$\bigcirc$	$\bigcirc$	$\bigcirc$	$\bullet$	$\bigcirc$
Analyzed in this work	$\bullet$	$\bullet$	$\bullet$	$\bullet$	$\bigcirc$	$\bullet$

2   CPS: Background and Related Work

CPS Architecture. Cyber-physical systems encompass a wide spectrum of applications [23]. The general CPS architecture consists of three main components. Sensors: measure the physical environment; controllers: use the information received from sensors and decide which actions to take; and actuators: execute those commands.

CPS Security. Given the high degree of interconnections in a CPS, the overall security of CPS deployments relies on trustworthy communication. In practice, CPS systems are often deployed relying on protocols that do not implement security features (such as authentication or encryption) e.g., Fieldbus [14], CAN [14], or mavlink [21]. Communication protocols that promise security were introduced for ICS, but in practice, there are challenges in deploying secure CPS [10].

Attacks to CPS. CPS are important for our society and they are a valuable target of attacks [6]. Attacks on CPS occurred in the past. For example, attacks to ICS and critical infrastructures e.g., Stuxnet [31] targeting nuclear plants, the Colonial Pipeline attack [32] targeting gasoline pipeline and Oldsmar's water treatment attack [7] targeting a water facility. The common goal of attacks is to physically or remotely exploit the CPS to cause process disruption.

Anomaly Detection for CPS. A number of process-based anomaly detection techniques were proposed in the literature. They leverage the characteristics of the physical process to detect deviations in the process data caused by attacks [6]. *i) Residual-based approaches* are trained to minimize a loss function (usually Mean Squared Error), between the expected and observed sensor readings. To detect anomalies, the loss between input and output is monitored, if it exceeds a threshold an alarm is raised. In this category, we find control theoretic approaches e.g., Auto Regressive (AR) models [19] and Linear Time-Invariant (LTI) models [9,29], and machine learning approaches e.g., Support Vector Machines [2,8] and Deep Neural Networks [16,22,28]. *ii) Invariant-based approaches* consist of rules that describe conditions that always hold in a given state on the CPS [1]. Those rules are often written based on detailed process knowledge [1,13].

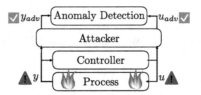

Fig. 1. System and attacker model, we assume a physical process that is controlled by a controller and monitored by an anomaly detection system. The attacker wants to hide an ongoing anomaly on the CPS. The attacker is aware that anomaly detection is deployed and wishes to conceal the true state and evade detection.

Evasion Attacks Against CPS. In Adversarial Machine Learning, evasion attack refers to the setting in which an attacker modifies a sample to induce misclassification in a classifier [4]. In the context of the Advanced Driver-Assistance System (ADAS), several attacks were proposed e.g., against LIDAR [5], location estimation [27]. In the context of CPS anomaly detection, white-box attacks against Deep Learning models [11,33] were demonstrated. Also, generic black-box evasion techniques were proposed [12]. Table 1 summarizes prior work in the field of anomaly detection for CPS, and reports which models were analyzed before for white-box concealment attacks. In this work, we focus on models proposed in prior work but not analyzed so far against white-box concealment.

3   System and Attacker Model

We assume a Cyber-Physical System that is monitored by an anomaly detection system to detect anomalies (Fig. 1). The physical process is controlled by one or multiple controllers, control commands u and sensor readings y are observed by the anomaly detector and used for the detection. Consistent with related work [11], we assume an attacker that has physical access to the CPS e.g., the attacker can attach malicious hardware to the network, and perform sensor spoofing exploiting communication protocol vulnerabilities (e.g., unauthenticated industrial protocols [14]) or performing attacks such as Man-in-the-PLC [15] attack. The attacker has knowledge of the system and can query the anomaly detector to obtain the predictions/classifications w.r.t. the current y and u. The attacker's goal is to launch a concealment attack to hide an ongoing process anomaly in the system (i.e., conceal the anomalies caused by the attacker on the process from the anomaly detection system).

The attacker can modify exchanged industrial traffic in transit, or compromise intermediate hosts to change values being forwarded (y_{adv} and u_{adv}), in Fig. 1. For example, in the Stuxnet attack [31] a compromised PLC was changing the rotation frequency of centrifuges of a nuclear process while reporting the correct frequency value to the anomaly detection to hide the anomaly. We measure the cost of the attack with respect to the number of features that are manipulated using the L0 norm (independent of the modification amount, i.e. L2 norm), as the effort is in compromising the communication channel, and at that point, arbitrary values can be set [11]. In practice, we allow any perturbations within the operational limits of the respective sensor or actuator [26].

3.1 Research Goals and Challenges

We address the three open research questions presented in the introduction. While addressing the three research questions we tackle the following research challenges: **C1** The attacker manipulates dynamic streaming data on the fly, which means that the attacker i) iteratively manipulates each value sequentially without knowing future values in advance; ii) adapts the strategy according to previous values stored in data logs without altering them. This is imposed by the Cyber-Physical Systems, where the attacker is assumed to perform sensor spoofing exploiting communication channels vulnerabilities. **C2** Not all general detectors are guaranteed to have differentiable loss functions (in contrast to Deep Learning based detectors). Thus, we need a general technique to attack different detectors even in absence of a loss function. For example, the detector [13] represents the current sample as a boolean vector (each element representing whether a specific invariant was violated). For this reason, we cannot use gradient-based methods (for example) to find optimal evasion samples.

Our main goal is to assess whether additional knowledge on detection mechanisms (i.e. white-box attacks) allows the attacker to perform better compared to black/grey-box attacks discussed in prior work [11,12]. This allows us to assess the robustness of CPS anomaly detectors, i.e., the minimal number of communication channels (features) that need to be controlled by the attacker to avoid detection.

3.2 Formal Definition of Concealment Attack

We now summarize the formal definition of the attack based on prior work [11]. Sensor and actuator values from a CPS are logged and used for anomaly detection. Given an anomalous feature vector $x = (y, u)$ (i.e., sensors and actuators readings) collected at a certain instant in time, a binary classification function $f(x)$ that classifies system state as 'anomalous' or 'safe', the concealment attack looks for a feature perturbation δ that added to x produces target misclassification (Eq. 1).

$$
\begin{aligned}
\text{Given} \quad & x = (y, u) \\
\text{s.t.} \quad & f(x) = \text{`anomalous'} \\
\text{Find} \quad & x_{adv} = x + \delta \\
\text{s.t.} \quad & f(x_{adv}) = \text{`safe'}
\end{aligned}
\tag{1}
$$

where $y \in \mathbb{R}^n$, $u \in \mathbb{R}^m$, $x \in \mathbb{R}^{n+m}$, $x_{adv} = (y_{adv}, u_{adv})$, $y_{adv} = y + \delta_y$, $\delta_y \in \mathbb{R}^n$, $u_{adv} = u + \delta_u$, and $\delta_u \in \mathbb{R}^m$.

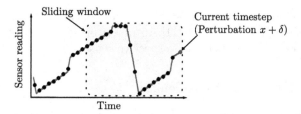

Fig. 2. Challenge **C1**. For each time slot, the attacker can only manipulate the latest sensor reading without knowing future values. We note the attacker cannot retroactively modify previous (manipulated or original) values. Eventually, the data considered in the sliding window will exclusively process values that were manipulated before.

4 Proposed Approach

We design a generic white-box concealment attack for CPS anomaly detectors. In this section, we start proposing the general framework that can be applied to attack prior work anomaly detectors.

4.1 White-Box Concealment Attacks (WBC)

We translate the white-box concealment attack (WBC) objective (Eq. 1) into an error minimization problem (Eq. 2)

$$\text{minimize} \quad Loss_{x_{adv}}(x_{adv}, tc)$$
$$\text{where} \quad tc = \text{target class} \tag{2}$$

Then, we induce targeted misclassification (to achieve the goal in Eq. 1) inspired by the Fast Gradient Signal Method (FGSM) [18] proposed originally for the domain of image manipulation.

$$\delta = -\epsilon * \text{sign}(\nabla_x Loss(x, tc)) \tag{3}$$

Every anomaly detection method has a different classification function, and consequently a different loss (if explicitly present), for this reason, this generic method is suitable to be applied to different categories of anomaly detectors.

The perturbation in Eq. 3 is iteratively applied until the concealment attack is successful and the detector no longer flags the anomaly (Eq. 1). Two attackers can be considered in this setting (we will compare them in Sect. 6). The first continue iterating until the objective (Eq. 2) is minimized, and the second continues until the classification label is changed, but the objective is not necessarily minimized.

4.2 Attacking Detectors with Differentiable Classifiers

We now address the research challenge **C1**: on-the-fly manipulation of streaming data (see Fig. 2). Residual-based anomaly detectors classify anomalies based

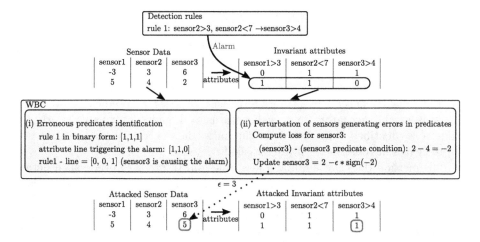

Fig. 3. Challenge **C2**. WBC concealment against detectors without loss function. Invariant-based detection works by checking whether the sensor readings satisfy the invariant rules, without relying on predictive models (no loss function). To apply our WBC attack, we re-formulate invariant-based methods as loss-based. We manipulate the sensors and actuators readings according to the difference (loss) between the desired state specified by the rules and the current state.

on the residual error between the sensors and actuators readings x and a predicted output value o from the anomaly detection classifier. That classification is performed over a sliding window of past observed values (i.e., $[x_{t-n}, \ldots, x_t]$).

In our scenario, the attacker can not simultaneously manipulate each value in this sliding window (as it would require post-hoc change of data), only the current sensor reading x_t can be manipulated. This introduces a novel constraint on the attack as the attacker has to minimize the residual loss acting on the last observed sample, and cannot globally minimize the loss function. We account for this additional constraint in our evaluation.

For example, to perform the WBC attack for residual-based detectors, we model the residuals by using the Mean Squared Error loss (Eq. 4). Then, we compute the partial derivative of the mean squared error w.r.t. x (Eq. 5) and apply directly FGSM to it (Eq. 3).

$$Loss(x,o) = \frac{1}{2}(x - o)^2 \quad (4) \qquad \nabla_x Loss(x,o) = x - o \quad (5)$$

4.3 Attacking Detectors with Non-differentiable Classifiers

Invariant-based anomaly detectors [1,13] classify anomalies based on the coherence of the system sensors and actuators w.r.t. a set of process invariant rules. When invariants are used, detectors check if some invariant rules are not fulfilled and raise an alarm consequently.

$$\text{Given an invariant rule } R: \quad A \rightarrow B \tag{6}$$
$$\text{(read as: if A then B)}$$

where A is the antecedent and B is the consequent of the invariant rule. Antecedent and consequent of a rule, consist of a set of predicates over certain sensors and actuators (e.g., valve_status $= 1$ and sensor_value < 4). An anomaly is identified if predicates in the antecedent A are all satisfied but not all predicates in consequent B are satisfied.

This method does not employ a loss function. In order to evade such detectors we need to consider the research challenge **C2**, i.e., we need to formulate the invariant-based approach as a loss-based method. Specifically, to evade the detector an attacker is required to modify the sensor readings in such a way that the predicates in B are fulfilled[1]. In order to do so, we decompose the attack in two steps (Fig. 3 provides a toy example of the method).

(i) Erroneous Predicates Identification. In the first step we identify which predicates trigger the anomaly in B. To do so we perform the set difference between the predicates in the rule R and the predicates observed in the system P (Eq. 7).

$$R \setminus P \tag{7}$$

Practically predicates are represented by Boolean conditions (i.e., boolean vectors where the position represents a certain invariant and the value 1 or 0 represents if the invariant condition holds). We identify the predicate that does not match the triggered rule performing the difference of such vectors.

(ii) Perturbation of Sensors Generating Errors in Predicates. In the second step, for the predicates that are erroneous we need to perturb the data related to that predicate to induce the change in the generated predicates. To guide sensor reading perturbation we can consider the desired value (i.e., the condition required by the predicate) of the erroneous predicates as our target value. This step can be performed by substituting the desired value directly in the sensor reading if the predicate is a direct equality or inequality over the sensor value (e.g., sensor $= 3$). Otherwise, if the predicate aggregates more information about a sensor reading (e.g., Gaussian Mixture Models over sensor value updates), we formulate the problem as a Mean Squared Error minimization as in Sect. 4.2, and compute the perturbation using Eq. 3 and using the loss as in Eq. 4.

5   Implementation and Evaluation Setup

In this section, we provide details about the implementation setup, the target anomaly detection systems, and the dataset used for evaluation. Based on the

---

[1] Alternatively the attacker can deactivate a rule by violating one condition in A, but this does not give guarantees about other rules that might be triggered by the modification.

categories of detectors identified in Sect. 2 and the analysis of prior work white-box concealment attacks in Table 1 we selected the target detectors according to three main criteria: (i) diversity of the detection technique (ii) not covered by prior work studies on white-box concealment attacks (iii) code availability for the detector. Our selection covers the research gap in the field of white-box concealment attacks on CPS anomaly detection. We consider five different anomaly detectors proposed in relevant prior peer-reviewed publications; namely, Auto Regressive model [19], Linear Time Invariants [29], Support Vector Machines [3,8], Process Invariants [13], and for each, we apply our proposed approach to achieve misclassification.

5.1 Attack Implementation and Hardware Setup

All experiments were performed on a laptop, equipped with Intel(R) Core(TM) i7-8650U CPU @ 1.90GHz, and 16GB of RAM. Experiments were performed either using Matlab 2019a, or Python 3.8.10 (depending on detector sources).

Implementation of the attack required: 201 lines of Matlab code for the AR model [19], 249 lines of Matlab code for the LTI model [29], 287 lines of Python code for the SVM [8] in this case we relied on the secml [25] library for gradients calculation by creating a wrapper for sklearn OneClassSVM, 324 lines of code for the PASAD detector [3], and 490 lines of code for the SFIG detector [13]. The code of our attacks is available at https://github.com/scy-phy/whiteboxDimva23.

5.2 Auto Regressive Models

AR models are a popular method used to model time series processes using linear equations starting from process data. Specifically, an Auto Regressive model (Eq. 8), tries to minimize the prediction error of sample X_t given the previous values $(X_0 \ldots X_{t-1})$.

$$X_t = c + \sum_{i=1}^{p} \gamma_i X_{t-i} + \epsilon_t \tag{8}$$

where c is a constant, $\gamma_i, \ldots, \gamma_p$ indicates the parameters of the model and ϵ_t is white noise. The parameters of the model are fitted using Yule-Walker equations [20]. AR models were applied to perform anomaly detection in cyber-physical systems [19,29]. The AR model is fitted starting from normal operations data, consequently, residuals observed during training are used to identify some thresholds or to tune Cumulative Sum (CUSUM) statistics. At test time the residuals are monitored to detect some deviations from expected behavior. **Availability**. We relied on the re-implementation by Erba et al. [12] and adapted it to work with SWaT dataset.

5.3   Linear Time Invariant Models

Linear Time Invariant (see Eq. 9) models were applied [9, 29].

$$\begin{cases} s_{k+1} = As_k + Bq_k \\ x_k = Cs_k + Dq_k \end{cases} \tag{9}$$

where $k := kT$ and T is the sampling time. $s_k \in \mathbb{R}^n$ is the state of the system, i.e., the variables (directly or indirectly observable) of the process. $q_k \in \mathbb{R}^p$ is the input to the system. $x_k \in \mathbb{R}^q$ is the output of the system. $A \in \mathbb{R}^{n \times n}$ is the state matrix, relates the state s_k and its update s_{k+1}. $B \in \mathbb{R}^{n \times p}$ is the input matrix, relates the system input q_k and the state update s_{k+1}. $C \in \mathbb{R}^{q \times n}$ is the output matrix, relates the state s_k and the measured output x_k. $D \in \mathbb{R}^{q \times p}$ is the feed-through matrix, relates q_k and x_k.

Similarly to the AR model system identification (n4sid algorithm [30]) is applied to identify the LTI model parameters. Then the CUSUM algorithm performs anomaly detection. We identified an order 4 LTI model for the SWaT dataset. We use 22 sensors as input of the system, and the 3 tank level sensors as the output of the model. **Availability.** We relied on the re-implementation by Erba et al. [12] and we adapted it to work with the SWaT dataset.

5.4   SVM

We implement the SVM model proposed by Chen et al. [8], the proposed SVM is trained on the water tank sensor readings (π, π') measured at d timesteps from each other. To apply their proposed method to the SWaT dataset which contains exclusively benign samples in the training set, we switched to one class SVM classifier. Following the guideline in the paper we performed a grid search to tune the parameters of the SVM. The resulting model is OneClassSVM with linear kernel, γ=0.01, ν=0.02. We also tuned the parameter d. With our experiments, we tested d = 1, 10, 100, and 1000 s and found the best performance at 1 s. We note that the simulator used in the original paper has a faster sampling rate (5 ms) than the actual SWaT testbed sampling rate (1 s). **Availability.** This detector was made available to us by the authors of [8] upon request. We adapted it to work with the SWaT testbed dataset (originally it was proposed for the SWaT simulator).

5.5   PASAD

The PASAD model proposed in the work by Aoudi et al. [3] is based on the idea of Singular Value Decomposition (SVD) [17]. PASAD uses the time series data and applies a sliding window to them. Using the sliding window data samples, PASAD identifies a projection subspace where normal operations (i.e., the training data) sensor readings are projected to. Normal operations form a cluster in the projection subspace. At test time, if anomalous sensor readings occur on the system, the data points will be projected far away from the cluster obtained during training. The distance from the center of the cluster is used as a criterion to detect anomalies. **Availability.** This detector is available online on GitHub[2].

5.6   SFIG

The Systematic Framework for Invariant Generation (SFIG) method proposed by Feng et al. [13], based on the idea of process invariants (see Sect. 2), proposes a method to automatically find invariant rules starting from process data. The rules are generated based on three sets of predicates: distribution driven predicates, event driven predicates, and categorical predicates. Distribution driven predicates are generated fitting a Gaussian mixture model of the system, while categorical predicates are generated according to actuator states. Finally, event driven predicates are generated by fitting some linear models to capture critical values that trigger changes in actuator states. To perform anomaly detection, at each time step, sensor readings are tested against all the rules in the collection of identified rules. If a rule is not fulfilled an alarm is raised. **Availability.** This detector is available online on GitHub[3].

5.7   SWaT Dataset

SWaT [24] is a water treatment testbed located at the Singapore University of Technology and Design. It consists of a six-stage process for water treatment. Those six stages are controlled by interconnected PLCs, connected to Human Machine Interfaces (HMIs), Supervisory Control and Data Acquisition (SCADA) workstation, and a Historian. The SWaT dataset is a collection of data from 11 days of operations; 7 days were collected during the system in normal operation while 4 days were collected while 41 attacks were launched on the system. We rely on this dataset as it is commonly used in related research, notably, it was used to evaluate all the detectors from prior work that we test in this work against WBC.

6   Evaluation Results

In this section, we present the results of our evaluation. To answer to **R1**, we applied the five aforementioned detection mechanisms to the SWaT dataset [24] and attacked them with the proposed WBC. To answer **R2**, we verify the computational runtime of the proposed approach and the cost of the perturbations. Finally, to answer to **R3**, the results of the WBC attack methodology are compared against the performance when no concealment was applied to the data, and against the black-box attacks for CPS detectors [12].

For our proposed WBC attack we consider three variants. Namely, *WBC baseline*, where the WBC attack is applied to every set of sensor readings labeled

---

[2] https://github.com/mikeliturbe/pasad.
[3] https://github.com/cfeng783/NDSS19_InvariantRuleAD.

Table 2. WBC Attack on the AR model trained over SWaT sensor LIT301 (used as reference in prior work [3]). The WBC attacks evade the anomaly detection system (see original recall vs. WBC recall). μ indicates the mean, and σ the standard deviation. N indicates how many rows were modified by the attack [†]Note: technically NaN as the metric divides by 0.

Data	Acc	F1	Prec	Rec	FPR	Elapsed (ms) μ	σ	Euclidean D. μ	σ	N
Original	0.797	0.254	0.227	0.288	0.134	–	–	–	–	–
Prior Work [12]										
Replay	0.775	0.088	0.086	0.091	0.131	–	–	13.592	48.322	541
Random R.	0.832	0.501	0.389	0.702	0.151	–	–	12.832	45.903	541
Stale	0.788	0.186	0.173	0.201	0.131	–	–	17.341	55.804	522
Our WBC										
baseline	0.858	(0)[†]	0.000	0.000	0.024	0.004	0.014	11.046	40.578	515
NTP	0.860	(0)[†]	0.000	0.000	0.022	249.36	148.77	2.081	24.563	74
NA	0.879	(0)[†]	0.000	0.000	0.001	171.11	71.7	5.092	30.485	258

as 'anomalous' as ground truth (i.e., the attacker is manipulating the physical process), regardless if they are detected as anomalous or not. In this setting, the attacker iterates until the objective (Eq. 2) is minimized. This is the same setting considered by the attacks proposed by Erba et al. [12], and we use it for comparison. *No True Positives (WBC NTP)*, in this setting the WBC is applied to every set of sensor readings labeled as 'anomalous', which is also detected as anomalous by the anomaly detection system (i.e., physical anomaly correctly detected by the anomaly detection system). Finally, we consider the *No Alarms (WBC NA)*, in this setting the WBC is applied to every set of sensor readings that are detected as anomalous by the anomaly detection system (i.e., conceal also false positives). In WBC NTP and WBC NA settings, the attacker iterates until the label is changed.

We note that since there is the white-box assumption on the target detector, the attacker is assumed to access the prediction of the detector. Moreover, since the physical process manipulations are under the control of the attacker, the attacker knows when the physical process anomaly is occurring on the system (i.e., 'anomalous' ground truth in the SWaT dataset).

Evaluation Metrics. To assess the impact of the attack on the detection capability of the classifier we consider the following metrics: Accuracy, F1 score, Precision, Recall, and False Positive Rate. In particular, the Recall score gives us information on how the attack is capable of concealing the true state of the system from the anomaly detector. Elapsed time is measured to assess the mean computational overhead required by the WBC attack. Specifically, we measure average the time required to compute an adversarial example. Finally, we measure the Euclidean distance (L2) between the original sample p and the

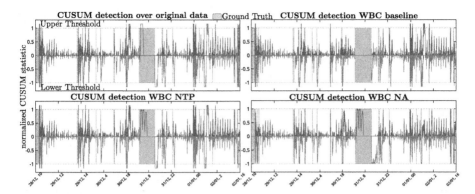

Fig. 4. Comparison of AR detection before and after the WBC attack. The concealment attack hides the anomalies in the process data. In the bottom figure (WBC NA) WBC is applied to all the readings even if no physical attack is present, this removes not only the True Positives but also the False Positives.

perturbed sample q to assess the perturbation required on the features by the attack. Moreover, to evaluate the minimal number of features under the control of the attacker we compute the Hamming distance (L0), as the number of sensors/actuators that were changed by the attack.

6.1 Auto Regressive

We apply the proposed approach to the AR detection model. In Table 2 we present the results of the WBC attack and compare them with the result from prior work black-box attacks [12], while Fig. 4 shows the impact of the WBC over the CUSUM statistics.

The AR detector precision and recall drop to 0 after the attack, this means that no more true positives are detected, and consequently, the F1 score becomes not defined as we have a division by zero. This result means that the detector is no longer capable of recognizing anomalies in the system. Looking at Fig. 4 we can also observe the difference between the three attack approaches (baseline, NTP, NA). WBC baseline brings the CUSUM error to 0 when the ground truth label reports 'anomalous', this happens because the attacker iterates until the loss is minimized. This is in contrast to WBC NTP and WBC NA for which the attacker stops iterating as soon as the alarm threshold is not surpassed anymore. Finally, we can notice the difference between the WBC NTP and WBC NA, the WBC NTP (as the name suggests) brings the True Positives to zero, while the WBC NA hides all the positives (both True Positive and False Positive).

Regarding the computational time, we observe that the WBC concealment attacks required hundreds of milliseconds to compute (while SWaT sampling time is 1 s). WBC baseline is sensibly faster because code optimization was used. As in the WBC baseline, we care of loss minimization, and we attack the AR model, we can achieve loss minimization in one step by selecting

Table 3. WBC Attack on the LTI model trained over SWaT.

Data	Acc	F1	Prec	Rec	FPR	Elapsed (ms) μ	σ	Euclidean D. μ	σ	N	Ham. D. μ	σ
Original	0.962	0.815	0.987	0.694	0.001	–	–	–	–	–	–	–
Prior Work [12]												
Replay	0.879	0.008	0.233	0.004	0.002	–	–	69.60	210.7	53863	21.4	1.8
Random R	0.998	0.992	0.987	0.996	0.002	–	–	69.58	210.53	53863	21.4	1.8
Stale	0.887	0.126	0.845	0.068	0.002	–	–	66.05	211.53	53862	19.8	4.1
Our WBC												
baseline	0.884	0.081	0.785	0.043	0.002	121.0	326.1	67.25	218.92	53863	2.9	0.3
NTP	0.885	0.087	0.831	0.046	0.001	32.9	22.9	59.41	208.37	37385	2.9	0.5
NA	0.885	0.087	0.944	0.046	0.000	87.8	46.4	59.89	208.95	37881	2.9	0.5

$\epsilon = ||\nabla_x Loss(x, o)||_2$ in Eq. 3. For clarity, this is equivalent to changing Eq. 3 to $\delta = -\nabla_x Loss(x, o)$.

We compare the white-box concealment technique w.r.t. the black-box attacks proposed by Erba et al. [12] (See Table 2). As we can observe the white-box attacks outperform the black-box attacks in terms of concealment capability, as the black-box attacks never conceal all the True Positives (i.e. recall greater than 0). Finally, we can compare the Euclidean distances between the attacks. As we can observe in Table 2, the average perturbation is always lower for the WBC attacks w.r.t. prior work black-box attacks. This is because the white-box setting optimizes the samples to be optimal w.r.t. the past observed process data. This is instead impossible for black-box attacks. This can be observed by looking at the number of modified values (N) in Table 2, which is always in favor of the WBC NTP ad NA attacks. Since the AR model is univariate, we do not report the hamming distance (it would be 1 in any case).

6.2   Linear Time Invariant

We apply the WBC concealment attacks to the LTI model. Table 3 reports the results of our evaluation. The WBC concealment attacks evade the LTI detector, and the detector recall drops from 0.69 to 0. We can observe the impact of the NA attack that reduces also the number of False Positive Rate.

The required computational time of the WBC attacks is at most 120 ms, which is lower than the sampling time of the SWaT system (1 s).

Also in this case the Euclidean Distance of the perturbed samples is lower than in prior work attacks. Moreover, when looking at the Hamming Distance, we can observe how the number of features to be manipulated decreases (2.93 vs 28.4). This happens because our WBC is constrained to manipulate the output of the model x_k but cannot operate on the input q_k (see Eq. 9). This number tells us that an attacker which controls 3 out of the 25 features used by the model, can significantly reduce the classifier recall by reducing it from 0.694 to 0.046.

Table 4. WBC Attack on the SVM model trained over SWaT sensor LIT101, LIT301, LIT 401.

Data	Acc	F1	Prec	Rec	FPR	Elapsed (ms) μ	σ	Euclidean D. μ	σ	N	Ham. D. μ	σ
Original	0.931	0.689	0.754	0.634	0.028	–	–	–	–	–	–	–
Prior Work [12]												
Replay	0.855	0.0	0.0	0.0	0.028	–	–	87.13	266.40	53897	5.99	0.12
Random R	0.855	0.0	0.0	0.0	0.028	–	–	87.45	266.03	53897	5.99	0.12
Stale	0.855	0.0	0.0	0.0	0.028	–	–	84.52	269.66	53896	5.84	0.55
Our WBC												
baseline	0.855	0.0	0.0	0.0	0.028	65.56	56.25	7.54	27.33	53934	5.99	0.07
NTP	0.855	0.0	0.0	0.0	0.028	103.84	33.13	7.54	27.33	34187	5.99	0.07
NA	0.880	0.0	0.0	0.0	0.000	107.22	61.49	10.71	36.74	45361	5.96	0.33

The number N is lower for WBC NTP and NA attacks when compared to prior work, i.e. 37881 vs 53863. When we compare the evasion performance of the attacks, we can observe that the WBC approach has comparable performance to the Replay and Stale attacks in terms of reduction of the model recall.

6.3 SVM

Table 4 reports the results of the evaluation of the proposed attacks on the SVM model. The proposed WBC concealment attack evades the SVM model and the recall drops from 0.63 to 0 in all the considered settings. We can observe how the NA approach differs by bringing the FPR to 0. The average computational time is at most 107 ms, which is lower than the sampling rate of the SWaT testbed. Since the detector is using the water tank levels measured at d timesteps of distance (π,π'), we constraint the adversarial example to modify only π' (3 features), this is consistent with our challenge C1.

When comparing the WBC attacks to prior work generic concealment attacks, we can observe that the Euclidean distance required by the WBC attack is lower, as well as the number of perturbed samples by the NTP and NA approaches. The Hamming distance remains almost the same, after d timestep of continuous attack (in our case 1 step) all the features in (π,π') are under the control of the attacker (6 features).

6.4 PASAD

In this section, we attack PASAD with our WBC approach. The results of the attack are summarized in Table 5. Also, in this case, the attacks are successful and the performance of the detector is compromised, as the recall drops close to 0 in all the three considered attacks. Differently from the previous case, the recall does not reach exactly 0, this is because there are a few instances in which the WBC is not reducing enough the distance from the PASAD cluster

Table 5. WBC results on PASAD trained on SWaT Dataset sensor LIT301 (used in the paper [3]). The WBC attacks evade PASAD. The WBC requires less than 4ms to compute. The Euclidean distance is smaller when compared to prior work attacks. Threshold 3×10^6.

						Elapsed (ms)		Euclidean D.		
Data	Acc	F1	Prec	Rec	FPR	μ	σ	μ	σ	N
Original	0.878	0.557	0.492	0.641	0.090	–	–	–	–	–
Prior Work [12]										
Replay	0.822	0.118	0.145	0.100	0.080	–	–	13.664	48.693	53859
Random R	0.819	0.083	0.106	0.069	0.079	–	–	13.341	47.829	53852
Stale	0.899	0.617	0.563	0.681	0.072	–	–	17.356	56.065	52013
Our WBC										
baseline	0.825	0.039	0.057	0.03	0.067	2.31	1.84	12.92	48.716	40962
NTP	0.818	0.008	0.011	0.006	0.072	3.91	7.95	8.265	94.526	24842
NA	0.870	0.004	0.025	0.002	0.012	2.6	2.44	10.431	48.302	33369

center. Similar to the previous experiment, we can see the difference between the baseline, NTP, and NA approaches. Again we can observe how the FPR rate reduces in the case of the NA setting. This time it reaches 0.012 meaning that there are few false positives.

Looking at the computational time required, the WBC algorithm finds the adversarial examples in 2.3 ms which is lower that the SWaT sampling time of 1 s. In this case, optimizations cannot be performed in the baseline setting. PASAD projects the univariate sensor readings into a subspace and tracks the distance of the projected time series form the centroid of the normal operations cluster. As explained with the research challenge C1, we assume we cannot change the whole time series sliding window but we manipulate just the last observation from the coming from the physical process. For this reason, the attack evades the detector by changing one sample at a time. Eventually, if the attack continues, all the samples in the sliding window are under the control of the attacker.

Finally, if we compare the performance of the white-box attacks w.r.t. black-box attack from prior work [12], we can observe that also in this case the WBC attacks are more effective than the black-box attacks in terms of concealment performance as the WBC recall score is always lower than in the case of the three attacks black box attacks from prior work. Looking at the Euclidean distance (Table 5), we can observe that the WBC attacks are on average less expensive than the black-box attack. Looking instead at the number of modified rows (N) we can observe that the WBC attacks are always less expensive than prior work.

6.5   SFIG

We then apply our attack method to the SFIG detector (see Table 6). In this setting, the WBC baseline and NTP coincide, because the invariant-based detector

Table 6. Attack against the SFIG detector on the SWaT dataset. The WBC baseline and NTP coincide because in invariant-based detectors alarms can be triggered only if rules are contradicted.

Data	Acc	F1	Prec	Rec	FPR	Elapsed (ms) μ	σ	Euclidean D. μ	σ	N	Ham. D. μ	σ
Original	0.958	0.793	0.950	0.681	0.005	–	-	–	–	–	–	–
Prior Work [12]												
Replay	0.876	0.000	0.004	0.000	0.005	–	–	69.60	210.70	53863	28.4	5.3
Random R	0.893	0.240	0.797	0.141	0.005	–	–	69.58	210.53	53863	28.4	5.3
Stale	0.881	0.080	0.544	0.043	0.005	–	–	66.05	211.53	53862	27.4	8.4
Our WBC												
base./NTP	0.876	0.003	0.040	0.002	0.005	256.2	34.5	0.136	0.459	36704	2.8	0.5
NA	0.880	(0)[†]	0.0	0.0	0.0	354.3	264.6	0.141	0.465	38643	2.8	0.6

triggers only when rules are contradicted (i.e., there is no loss to minimize). In this experiment, we consider attacks that deal with the 51 features of the SWaT dataset, as the detector considers them all together.

Also in this setting, the detector was evaded by the attacks reducing the performance of the detector from 0.68 to 0 in both cases. Also here we can appreciate the difference induced in the false positive rate in the two attack settings, the NA setting leaves no false positives.

In the WBC baseline/NTP, we notice that the recall is not 0.000, this is because we noticed that there is an artifact in the detection rules which causes a contradictory set of rules. This means that applying our attack to fix the data to turn off the alarms, triggers another rule in contradiction. This makes it impossible to turn off the alarm in a row of data.

Looking at the computational time required by the attacks in this case, we are in the order of 200/300 ms which is lower than the SWaT sampling time. Regarding the Euclidean distance, from Table 6 we can observe that the WBC attacks are less expensive than the attacks from the black box attacks from prior work, the proposed WBC attacks are always 2 orders of magnitude closer to the original values, meaning that features need to be slightly modified to achieve the goal. Also, the number of modified rows N (as in the previous experiments) is smaller. In this multivariate setting, we can also measure the number of features that were modified by the attack (i.e., the Hamming distance). As we can observe in Table 6, out of the 51 features in the SWaT dataset, WBC attacks modify on average 2.8 features (maximum 7 features out of 51), while prior work attacks modify on average ∼30 features (maximum 37 features out of 51).

Finally, if we compare the performance of the WBC w.r.t. attacks from prior work [12], we can observe that on one hand, the WBC NTP have a similar performance to Replay and Stale attacks from prior work, but on the other hand, as we pointed out before the WBC NTP is overall cheaper in terms of features that are modified by the attack.

Table 7. Summary of findings on our white-box concealment attacks. '# Manipulated' refers to the number of features that needed to be manipulated by the attacker.

Method	Attack works	# Manipulated	Computational Cost ≤ 1 s
AR	✓	1/1	249 ms
LTI	✓	3/25	120 ms
SVM	✓	3/6	107 ms
PASAD	✓	1/1	4 ms
SFIG	✓	3/51	360 ms

7   Discussion and Conclusion

In this section, we discuss the answers to our research questions. In Table 7 we summarize our findings. With respect to question **R1**, we tested three variations of the proposed WBC attacks, over five different anomaly detection systems. To do so the attacker has to deal with challenge **C1** (i.e., manipulate only the last sensor value) and with challenge **C2** (i.e., transform to differentiable detectors which do not use a loss function). As a result, we found that the evaluated detectors are vulnerable to white-box concealment attacks, i.e., for all the tested detectors, the recall score drops to 0 or very close to it. This result demonstrates that the proposed attack methodology can affect a wide range of anomaly detectors for cyber-physical systems, affecting their detection performance with often little perturbation of the sensor data (in terms of Hamming and Euclidean distance). Our analysis reveals that only a low number of resources need to be under the control of an attacker to subvert the classification outcome of the target anomaly detector. For example, for the LTI and the SFIG, our results show that is enough to control ∼3 features of the multivariate detector to conceal attacks.

With respect to research question **R2**, we measured the time to compute the adversarial examples (worst case ∼350 ms), and we found that runtime manipulations are possible, as it is possible to compute manipulations faster than the system's sampling rate of the SWaT system (1 sample per second). We note that temporal constraints for adversarial examples are not generally investigated by related adversarial machine learning literature, as in other domains adversarial examples can be pre-computed (for example in the image classification domain) and do not need to be adapted based on the context.

Concerning research question **R3**, we compared the proposed attacks with black-box attacks from prior work [12], in particular in terms of concealment performance and Euclidean distance. We found that our proposed WBC attacks are more effective (e.g., F1 score of the PASAD model is always lower in the WBC attack 0.039 vs 0.083 from prior work). Moreover, in general, our attacks require less manipulation than prior work attacks, (e.g., the Euclidean Distance in the SFIG case is 0.136 vs 66.05 from prior work, same the holds for the Hamming distance WBC 2.81 vs 27.41 from prior work).

Our results demonstrate that it is possible to evade a wide range of detectors while reducing the number of samples that need to be manipulated (compared to prior black-box concealment attacks). Those findings highlight the need for further research and constructive discussion about guarantees for CPS anomaly detectors against adversarial manipulation. As such we see our contribution toward the robustness and reliability of CPS detectors against adversarial examples.

References

1. Adepu, S., Mathur, A.: Distributed detection of single-stage multipoint cyber attacks in a water treatment plant. In: Proceedings of the ACM ASIA Conference on Computer and Communications Security (ASIACCS) (2016)
2. Ahmed, C.M., et al.: Noiseprint: attack detection using sensor and process noise fingerprint in cyber physical systems. In: Proceedings of the Asia Conference on Computer and Communications Security (AsiaCCS) (2018)
3. Aoudi, W., Iturbe, M., Almgren, M.: Truth will out: departure-based process-level detection of stealthy attacks on control systems. In: Proceedings of the ACM Conference on Computer and Communications Security (CCS). ACM (2018)
4. Biggio, B., et al.: Evasion attacks against machine learning at test time. In: Blockeel, H., Kersting, K., Nijssen, S., Železný, F. (eds.) Machine Learning and Knowledge Discovery in Databases, pp. 387–402 (2013)
5. Cao, Y., et al.: Adversarial sensor attack on lidar-based perception in autonomous driving. In: Proceedings of the ACM Conference on Computer and Communications Security (CCS), p. 2267–2281. ACM, New York, NY, USA (2019)
6. Cárdenas, A., Amin, S., Sinopoli, B., Giani, A., Perrig, A., Sastry, S.S.: Challenges for securing cyber physical systems. In: Workshop on Future Directions in Cyberphysical Systems Security. DHS, July 2009
7. Cervini, J., Rubin, A., Watkins, L.: Don't drink the cyber: extrapolating the possibilities of Oldsmar's water treatment cyberattack. In: International Conference on Cyber Warfare and Security, vol. 17, pp. 19–25 (2022)
8. Chen, Y., Poskitt, C.M., Sun, J.: Learning from mutants: using code mutation to learn and monitor invariants of a cyber-physical system. In: Proceedings of the IEEE Symposium on Security and Privacy, pp. 648–660. IEEE (2018)
9. Choi, H., et al.: Detecting attacks against robotic vehicles: a control invariant approach. In: Proceedings of the ACM Conference on Computer and Communications Security (CCS) (2018)
10. Dahlmanns, M., Lohmöller, J., Fink, I.B., Pennekamp, J., Wehrle, K., Henze, M.: Easing the conscience with OPC UA: an internet-wide study on insecure deployments. In: Proceedings of the ACM Internet Measurement Conference (2020)
11. Erba, A., et al.: Constrained concealment attacks against reconstruction-based anomaly detectors in industrial control systems. In: Proceedings of the Annual Computer Security Applications Conference (ACSAC), December 2020
12. Erba, A., Tippenhauer, N.O.: Assessing model-free anomaly detection in industrial control systems against generic concealment attacks. In: Proceedings of the Annual Computer Security Applications Conference (ACSAC). Austin, USA, December 2022

13. Feng, C., Palleti, V.R., Mathur, A., Chana, D.: A systematic framework to generate invariants for anomaly detection in industrial control systems. In: Proceedings of Network and Distributed System Security Symposium (NDSS) (2019)
14. Galloway, B., Hancke, G.P., et al.: Introduction to industrial control networks. IEEE Commun. Surv. Tutor. **15**(2), 860–880 (2013)
15. Garcia, L., Brasser, F., Cintuglu, M.H., Sadeghi, A.R., Mohammed, O., Zonouz, S.A.: Hey, my malware knows physics! attacking PLCs with physical model aware rootkit. In: Proceedings of Network and Distributed System Security Symposium (NDSS), February 2017
16. Goh, J., Adepu, S., Tan, M., Lee, Z.S.: Anomaly detection in cyber physical systems using recurrent neural networks. In: 2017 IEEE 18th International Symposium on High Assurance Systems Engineering (HASE), pp. 140–145. IEEE (2017)
17. Golub, G.H., Reinsch, C.: Singular value decomposition and least squares solutions. In: Bauer, F.L. (eds.) Linear Algebra, vol. 2, pp. 134–151. Springer, Heidelberg (1971). https://doi.org/10.1007/978-3-662-39778-7_10
18. Goodfellow, I.J., Shlens, J., Szegedy, C.: Explaining and harnessing adversarial examples. CoRR abs/1412.6572 (2014)
19. Hadžiosmanović, D., Sommer, R., Zambon, E., Hartel, P.H.: Through the eye of the plc: semantic security monitoring for industrial processes. In: Proceedings of the Annual Computer Security Applications Conference (ACSAC), pp. 126–135. ACM, New York, NY, USA (2014)
20. Hayes, M.H.: Statistical Digital Signal Processing and Modeling. Wiley, Hoboken (2009)
21. Koubâa, A., Allouch, A., Alajlan, M., Javed, Y., Belghith, A., Khalgui, M.: Micro air vehicle link (MAVlink) in a nutshell: a survey. IEEE Access **7** (2019)
22. Kravchik, M., Shabtai, A.: Detecting cyber attacks in industrial control systems using convolutional neural networks. In: Proceedings of the 2018 Workshop on Cyber-Physical Systems Security and PrivaCy, pp. 72–83. ACM (2018)
23. Lee, E.A.: Cyber physical systems: Design challenges. Technical report UCB/EECS-2008-8, EECS Department, University of California, Berkeley, January 2008
24. Mathur, A., Tippenhauer, N.O.: SWaT: a water treatment testbed for research and training on ICS security. In: Proceedings of Workshop on Cyber-Physical Systems for Smart Water Networks (CySWater), April 2016
25. Melis, M., Demontis, A., Pintor, M., Sotgiu, A., Biggio, B.: secML: a Python library for secure and explainable machine learning. arXiv:1912.10013 (2019)
26. Pierazzi, F., Pendlebury, F., Cortellazzi, J., Cavallaro, L.: Intriguing properties of adversarial ML attacks in the problem space. In: Proceedings of the IEEE Symposium on Security and Privacy, pp. 1332–1349. IEEE (2020)
27. Shen, J., Won, J.Y., Chen, Z., Chen, Q.A.: Drift with devil: security of multi-sensor fusion based localization in high-level autonomous driving under GPS spoofing. In: Proceedings of the USENIX Security Symposium, pp. 931–948, August 2020
28. Taormina, R., Galelli, S.: A deep learning approach for the detection and localization of cyber-physical attacks on water distribution systems. J. Water Resourc. Plann. Manag. **144**(10) (2018)
29. Urbina, D., et al.: Limiting the impact of stealthy attacks on industrial control systems. In: Proceedings of the ACM Conference on Computer and Communications Security (CCS), October 2016
30. Van Overschee, P., De Moor, B.: N4sid: subspace algorithms for the identification of combined deterministic-stochastic systems. Automatica **30**(1), 75–93 (1994)

31. Weinberger, S.: Computer security: is this the start of cyberwarfare? Nature **174**, 142–145 (2011)
32. Wikipedia, t.f.e.: Colonial pipeline ransomware attack. https://en.wikipedia.org/wiki/Colonial_Pipeline_ransomware_attack. Accessed 21 May 2022
33. Zizzo, G., Hankin, C., Maffeis, S., Jones, K.: Adversarial attacks on time-series intrusion detection for industrial control systems. In: IEEE TrustCom (2020)

A Security Analysis of CNC Machines in Industry 4.0

Marco Balduzzi[1](✉), Francesco Sortino[2], Fabio Castello[2],
and Leandro Pierguidi[2]

[1] Trend Micro Inc., 225 East John Carpenter Freeway, Irving, Texas, USA
marco.balduzzi@madlab.it

[2] Celada SpA, via Cesare Battisti 156, Cologno Monzese, Milan, Italy

Abstract. Computer numerical control (CNC) machines are extensively used in production plants and are considered a crucial asset for organizations worldwide. These machines require unique controllers that differ from those used in other types of machine tools in terms of software architecture, protocols, and design, so to meet the high precision and accuracy demands of their applications. The growing adoption of network-enabled systems in the industrial domain, driven by Industry 4.0, has resulted in an increased use of CNC machines. These machines have evolved from traditional mechanical machines to full-fledged systems with multiple networking services for smart connectivity. This study investigates the risks associated with this technological development. Using actual machine installations, we conducted the first empirical evaluation of the privacy and security implications of Industry 4.0 in the CNC domain. Our findings revealed that malicious users could conduct five types of attacks: compromise, denial-of-service, damage, hijacking, and theft. We reported our findings to the affected vendors and proposed mitigations to manufacturers, integrators and end-users. Our work aims to provide an opportunity to increase awareness in a domain where security does not appear to be a priority at present.

1 Introduction

In the past decade, there has been a significant rise in the popularity of network-enabled systems, even for devices that were historically not designed to offer such capabilities. This trend has been particularly evident in the industrial domain, where various types of network-enabled systems are widely used to support modern manufacturing processes.

The development of devices such as industrial gateways, computer numerical controls (CNCs), industrial robots, and autonomous vehicles for logistics has led to new industrial models that follow the general paradigm of Industry 4.0, driving manufacturing companies towards networked shop floors. While connecting modern machine tools to wide networks, including the Internet, presents an important opportunity for creating new business intelligence through the collection and analysis of production data, it also poses potential threats to the security and privacy of organizations.

D. Gruss et al. (Eds.): DIMVA 2023, LNCS 13959, pp. 132–152, 2023.
https://doi.org/10.1007/978-3-031-35504-2_7

CNC machines play a fundamental role in the manufacturing industry because they are the building blocks of the mechanical processing of pieces. In a manufacturing line, a variety of systems cooperate, such as robots or other support systems (like control servers), but CNCs are responsible for the mechanical processing of the pieces through drillers, lathes, or cutters. Industrial robots, on the other hand, are used for auxiliary operations such as material handling, palletizing, or as soldering stations.

CNC machines require unique systems that differ from other machine tools, not only in terms of software architecture and protocols but also in their overall design, to meet the specific demands of precision and accuracy required by their applications. While under the hood, CNC machines still rely on well-established mechanical automation routines, they are also equipped with unique solutions specific to their domain, such as advanced software algorithms and specialized hardware components. These domain-specific functionalities set modern CNCs apart from traditional machine tools and enable them to achieve higher levels efficiency in manufacturing processes.

For this reason, we believe that CNC machines are a key element in analyzing the security posture of the manufacturing ecosystem. As far as we know, we are the first to conduct a comprehensive analysis of the security issues related to this specific technology and demonstrate potential vulnerabilities in practice.

In short, the contributions of our work consist of the following:

- We investigate the security and privacy of CNC machines in Industry 4.0. To the best of our knowledge, we are the first to conduct a depth empirical analysis in this direction.
- We conduct an extensive security assessment of the technologies offered by modern CNCs by making use of the controllers provided by four large representative vendors.
- We perform threat modelling and report problems resulting in five attack classes: compromise, damage, denial-of-service, hijacking and theft.
- We communicate our findings to the affected vendors, we propose mitigations, and do our best to raise awareness in this domain.

2 Background

A CNC machine is a machine tool developed to transform the geometry of raw material through machining, a process through which a material (be it metal, polymer, or otherwise) is cut until it reaches the desired geometry through a *controlled process*. This process is carried out through the aid of cutting tools and is achieved using a controller, which, together with the mechanics of the machinery, constitutes a numerically controlled machine tool. The main benefit of this addition lies in the possibility of the machine to operate process phases in an unattended way and to use the computing power of the controller to create complex geometries with high degrees of precision.

CNC machines are programmed in G-code (RS-274 [1]). This language resembles Basic programming: it is presented as a series of instructions initialized by

(a) A Haas controller simulator.

(b) A Yasda machine running on Fanuc controller.

Fig. 1. Examples of simulator and machine.

a letter address, which follow one another on successive lines separated by paragraph breaks; each of these lines is called block. Each letter address specifies the type of movement or function called by the user in that part of the program. Over the years, concepts that we now consider basic in programming languages, such as loops, macros, and object programming, have also been introduced in machine language, and numerous examples of conversational or guided languages have been included to facilitate CNC operations.

While G-code is, still, the standard for programming CNC machines, engineers nowadays tend to rely on CAM software[1] to translate architectural drawings (of the parts to be produced) into software programs. Such programs are then ran on controller simulators before being deployed in production lines. Figure 1a shows one of these simulators, which can be either physical (like in the photo) or software (e.g., in form of virtual machine). Despite this difference, controller simulators implement the same logic of a real-world CNC machine (ref. Fig. 1b) – in fact, the software running on such simulators is normally the same as the one on the machine, despite the hardware peripherals being virtualized e.g. the motors used to move the machine's axes.

3   Approach

The manufacturing and deployment of a CNC machine can be modeled as a supply chain process, where a controller manufacturer produces and sells controllers to multiple machine manufacturers. The machine manufacturers, using

---

[1] Computer Aided Manufacturing.

the controllers, develop CNC machines such as lathes, and make them available to resellers, integrators, and end-users.

There are two considerations to make: firstly, any security issues or vulnerabilities introduced by the controller manufacturer at the beginning of the supply chain will be propagated throughout the entire chain, along with any technologies or software used by the controller. Therefore, by examining the controller, we can gain a wider perspective on the adoption of such technologies and any related issues throughout the supply chain.

Secondly, the number of controller manufacturers on the market is much smaller than the number of machine manufacturers, with a single controller typically being used to build dozens of CNC machines. This is important for our goal of evaluating the security of CNCs, as it means we can focus on a smaller number of manufacturers that represent a significant portion of the market.

Our investigation begins by identifying a set of representative, large controller manufacturers on the market. We proceed by selecting those players that have a worldwide reach, are on the market since tens of years, are widely known in the industrial domain, or have developed technologies widely used in this industry. All selected manufacturers develop controllers used on machines we have access to[2]. This is important for us because we want to conduct an empirical study, showing that our concerns have practical implications. Table 1 provides a summary of the selected manufacturers and their respective controllers and machines that we used for testing.

Our analysis consists of the following process:

- We conduct threat modelling, by presenting the scenarios in which a miscreant would be able to target a CNC machine and discussing the impact of such attacks.
- We identify the technologies introduced in the CNC realm to adhere to Industry 4.0. They encompass protocols and services used to connect the machines to smart environments, for example to share the production information with centralized systems for better management and cost reduction. They also enable remote management, for example, for an operator to change the executed program or configure the tooling.
- We conduct a first coarse-grained security assessment, for example using vulnerability scanners to identify potential known vulnerabilities or misconfigurations in such services. Note that the focus of our research is on domain-specific technologies, i.e. we ignore those problems related to generic software (like Windows services).
- We then go deep into the CNC technologies previously identified, by analyzing the risks of abuses and conducting practical attacks on the controllers. For this, we develop attack tools that leverage the weaknesses that we identified. We make use of both proprietary documentation and APIs we were given access to.

---

[2] The machines are located in different facilities: in Celada, MADE Competence Center, or the Department of Mechanical Engineering of the Polytechnic University of Milan.

Table 1. A summary of the selected manufacturers and their respective controllers and machines used for testing.

Vendor	Haas	Okuma	Heidenhain	Fanuc
Country	US	Japan	Germany	Japan
Year of establishment	1983	1898	1889	1972
Estimated size	More than US$1B revenue and 1,300 employees (2018)	US$1.41B revenue and 3,812 employees (2020)	US$1.3B revenue and 8,600 employees (2020)	US$4.18B revenue and 8,260 employees (2020)
Market	Controllers and machines for all markets	Controllers and machines for all markets	Controllers	Controllers and simple machines
Simulator	100.19.100.1123	OSP-P300S	TNC 640 Programming Station 340595 V.10.00.04	Not used
Controllers	100.20.000.1110	P300MA-H	TNC 640	31iB5 iHMI and 32i-B
Machines	Super Mini Mill	Genos M460V-5AX6	Hartford 5A-65E	Yasda YMC 430+RT10 and Star SR-32JII
Types	3-axis vertical machining center	5-axis vertical machining center	5-axis vertical machining center	5-axis vertical micro machining center and Swiss lathe

– We collect evidence of our concerns and collaborate with the affected vendors in suggesting mitigations.

3.1 Threat Modelling

CNC machines are commonly installed in manufacturing networks. These networks, often referred as OT networks, are standalone networks that traditionally were not in communication with corporate (IT) networks. However, in modern factory plants, CNC machines communicate with external servers for enabling remote machine programming or process monitoring. These machines are, for example located in corporate networks reachable via industrial gateways or mobile networks. Mobile operators offer connectivity to CNC machines via Internet while industrial gateways act as bridges between OT and IT networks. To confirm these trends, in the preliminary phase of our research, we conducted an interview with experts on the fields (e.g. suppliers and installers of machines) who confirmed these claims.

We model the attacker as following:

– A remote attacker who has access to the OT network. This attacker could be an insider with direct access to the OT network where the CNC machine is installed, or an attacker with a presence in an enterprise with missing or wrongly configured network segmentation that exposes the CNC machine.

- A remote attacker with access to the IT network. The attacker gains access to the CNC machine by pivoting from the IT network, potentially exploiting misconfigurations or vulnerabilities in the industrial gateway connecting the IT and OT networks. Previous research has shown that such devices are vulnerable to several types of attacks [2]. Alternatively, the attacker could pivot from the server that communicates with the CNC machine.
- An Internet-based attacker. In this scenario, the attacker conducts the attack from the Internet. Unfortunately, CNC machines are sometimes left exposed to the Internet for remote monitoring or due to misconfiguration. We conducted an analysis of this type using a large-scale scanner (ZMap) and found evidence of exposed machines. However, we did not connect to these machines for ethical reasons.
- A remote attacker who communicates with the machine operator. In this scenario, the attacker social-engineers the operator, for example, via email, persuading him to install a CNC add-in, as we discuss later.

In this threat model, we should also consider the possibility of an attacker with physical access to the machine. However, for the purposes of our research, we chose to focus solely on remote attackers and did not include this particular scenario.

An attacker who fits within our threat model would be capable of carrying out all five attack classes outlined in the rest of the paper.

3.2   CNC Technologies and Related Problems

All the controllers we considered provide various technologies that can integrate CNCs into modern digital shop floors. These technologies enable automatic data exchange with acquisition systems, enterprise resource planning (ERP) systems, CAM software, digital twin solutions, and tool management systems. Since these technologies are typically proprietary and designed specifically for CNCs, they require a thorough and specialized analysis to fully understand their security implications.

For example, Haas Connect[3] is a cloud service offered by Haas to monitor a machine remotely. With Haas Connect, an engineer can monitor the production information of the machine, knowing how many parts are produced over time, or being informed if any alert occurs. Many of these technologies are included by default in the controller, while others are offered on demand and need to be purchased in addition. However, we observed that most of the customers prefer purchasing machines equipped with all technologies for many reasons like the fiscal incentives offered by several countries on buying these "smart technologies" or the clear advantages in having machines that can be centrally managed and monitored. In our research, we decided to focus on those technologies that are included by-default in the installations (second column), with the addition of THINC-API for the reasons explained later. OPC-UA was not taken into consideration because rarely available.

---

[3] https://www.haascnc.com/productivity/control/haas-connect.html.

Table 2. A summary of Industry 4.0 technologies adopted by manufacturers.

Vendor	Default Technologies	Optional Technologies
Haas	MTConnect, Haas Connect, Ethernet Q Commands	NaN
Okuma	NaN	THINC-API , MTConnect
Heidenhain	RPC and LSV2 (DNC)	OPC-UA
Fanuc	Focas	OPC-UA , MTConnect

MTConnect[4] is an effort to standardize the different protocols used in the industrial domain to collect machinery data. The goal is indeed to provide guidelines for converting old and proprietary information to a common language; this will help organizations to handle machine tools from different brands in an easier form. Along with our evaluation, we confirmed that 3 of the tested vendors support MTConnect, in particular Haas provides such feature on all default installations. In our analysis, we investigated the data that an attacker could infer (or leak) from a machine exposing MTConnect over the network. A common scenario is, for example, the number of parts that are produced, together with the associated program. In other cases, an attacker can infer the source code of the executed program by repeatedly querying the MTConnect agent installed on the machine as we show later.

Despite the standardization effort around MTConnect, proprietary protocols are confirmed to be the majority, with one of these being Haas's Ethernet Q Commands[5]. With this protocol, a user can query information from a controller (for example the machine's model, the tooling configuration, or the number of produced parts) or set (program) variables needed for a program to execute. In the following Listing, few examples are given:

```
?100: Query the Machine's Serial Number
?Q402: Query the Parts Counter #1 (number of produced parts)
?Q600 10000: Read the value of variable 10000
?E10000 123: Write the value 123 into the variable 10000
```

This service is useful in making a machine reachable remotely and enables manufacturing automation; however, it may also expose the machine to potential threats. This is, indeed, the case suggested by our analysis. In fact, even if the documentation reported that only a limited range of registers could be written, namely those ones related to program variables (i.e., 10000-10999), this was not the case. As we describe later in the paper, our experiment confirmed that such a lack of access control allows a miscreant to conduct attacks like denial-of-service, hijacking, or damage.

Heidenhain offers so-called DNC interface[6], which is implemented with two protocols: RPC and LSV2. The first is a proprietary protocol operating on

---

[4] https://www.mtconnect.org.

[5] https://www.haascnc.com/service/troubleshooting-and-how-to/how-to/machine-data-collection---ngc.html.

[6] https://www.heidenhain.com/products/digital-shop-floor/connected-machining.

TCP/19003. Heidenhain uses the generic name of RPC (remote procedure call) for a protocol allowing a remote peer to call a remote interface's method on the CNC. The second is a standardized protocol used by certain vendors. While it is not as famous as other technologies, it is used and documented to a certain extent. PyLSV2 is, for example, a Python library for implementing a LSV2 compatible client.

In our evaluation, we obtained access to the RemoTools library provided by the manufacturer to the integrators in order to develop interfaces for the controller. A miscreant having access to this library is facilitated in implementing a malicious client for hijacking the operation of the CNC machine, or stealing confidential data. Note that the same attacks could be developed with public libraries as well, for example for LSV2. The controller offers the possibility to enable network authentication on the DNC interface for both RPC and LSV2. The authentication is implemented in form of SSH tunneling, which is very convenient because the controller runs on top of Linux. This option, which needs to be voluntarily enabled by the integrator or the end-user, is a good solution to the problems that we identified and that we discuss later in the paper.

Fanuc offers an equivalent technology called Focas[7]. Even though Focas offers a restricted set of remote-call possibilities compared with the other vendors (that is, a limited number of management features), our experiments showed that a miscreant can still conduct attacks like damage, DoS, and hijacking. This is an important issue because, unfortunately, authentication was introduced only recently (in 2020) and only as a non-default option - according to our communications with the vendor. This new version allows to configure an eight-digit code to be used as authentication token. This is achieved by setting the controller's global parameter 10344 to the desired code. By default, this value is set to 0 (no authentication).

Okuma stands out from the controller market for one interesting feature: the modularity of its controller. In fact, while the vendor offers in its simplest form a limited controller, it also provides a mechanism (called THINC-API) to highly customize its functionalities. With this technology, anyone can implement an add-in that - once installed - runs in the context of the controller, in the form of extension. Applications developed with THINC-API are commonly offered by integrators and resellers to their customers, and can be made available to 3rd-parties via the Okuma's app store[8] for easier distribution.

Given the prevalence of this technology, we conducted a dedicated assessment in the hope to better understand the potential impact of this technology despite not being provided as default option. Unfortunately, our analysis highlighted that simple security mechanisms that are nowadays very common like *resource access control* are not yet supported. As a result, if a miscreant manages to install a malicious application, she will be able to access all controller's information and to tamper with its behavior. There are several paths that a miscreant can take

---

[7] https://www.fanuc.eu/it/en/cnc/development-software/focas-development-libraries.

[8] https://www.myokuma.com/.

for such installation, for example by compromising the machine or using social engineering techniques. A malicious user could also upload the application to the app store, for example by hiding the malicious functionalities around legitimate ones, and lure her victim to download and install it. Note that we did not conduct this experiment for legal reasons. In our experiments, we managed to compromise the controller under test via a well-known system vulnerability (MS10-61) so as to install our application without notice. The malicious application we developed for testing mimicked a bot reaching out to the attacker via a call-back, and waiting for commands to be prompted to the backdoored CNC.

4   Findings

Our research reported issues common to many of the controllers under exam. We provide the summary of our findings and discuss their security implications.

First of all, the controllers we analyzed are equipped with either obsolete and legacy software, or software encompassing a large number of known vulnerabilities. Although this issue is well-understood in the ICS realm, and we were not surprised to run into obsolete software, we would have expected that machine tools like CNCs – that can easily cost a million dollar – would come with auto-updating mechanisms or, at least, mechanisms to inform the end-user of a need for an update. This is especially true in the context of Industry 4.0, in which machines tend to be normally connected to the network.

Second, several networking technologies do not support authentication, or do *not* have authentication enabled by default. In particular, only DNC and Focas have support for authentication, while MTConnect, Ethernet Q and THINC-API not have (note that THINC-API is a corner case because is exposed only locally). This issue is very severe because offers to any malicious user the possibility to abuse of the unauthenticated services.

Third, resource access control is lacking on most of the architectures of the controllers: A user (or a process) is often given full access to any system's resource, including its file-system or memory locations. For example, an application written on top of THINC-API will have full access permission to any system's resource including the internal controller configurations; with Ethernet Q, a remote user can write to memory locations mapped outside of the running process.

Fourth, the monitoring services expose a large amount of information. On one side, this is expected because those services have been, as said, introduced to make CNC machines compliant with Industry 4.0 paradigm. However, the information can be abused by a miscreant, especially given that authentication is often not available. In our experiments, we confirmed that all analyzed controllers suffer from data leakage problems resulting in confidential information being exposed to 3rd parties (e.g. programs code).

Table 3. Summary of the attacks identified in our research.

Attack Class	Attack Name	Haas	Okuma	Heidenhain	Fanuc	Total
Compromise	RCE	✓	✓	✓		3
Damage	Disable feed hold	✓				1
	Disable single step	✓		✓		2
	Increase tool life	✓	✓	✓		3
	Increase tool load	✓	✓		✓	3
	Change tool geometry	✓	✓	✓	✓	4
	Decrease tool life	✓	✓	✓		3
DoS	Decrease tool load	✓	✓		✓	3
	Change tool geometry	✓	✓	✓	✓	4
	DoS via parametric program	✓	✓	✓	✓	4
	Trigger custom alarms	✓		✓		2
	Ransomware	✓	✓	✓		3
Hijacking	Change tool geometry	✓	✓	✓	✓	4
	Hijack parametric program	✓	✓	✓	✓	4
	Program rewrite		✓	✓	✓	3
Theft	Leak production information	✓	✓	✓	✓	4
	Leak program code		✓	✓	✓	3
	Screenshot			✓		1
	Total	15	14	15	10	

4.1   Impact

Overall, as depicted in Table 3, our evaluation identified 18 attacks (or attack variations) that we grouped into five attack classes: compromise, damage, denial-of-service, hijacking, and theft[9].

Among the different controllers that we tested, we observed a consistency in the number of problems: Haas, Okuma and Heidenhain yielded a similar amount of issues (15), with Fanuc having 10 attacks confirmed. This is a symptom that security does not seem to be a priority for controller manufacturers. This, together with the possibility of CNC machines being misconfigured and exposed to corporate networks, or worse to the Internet, creates serious and compelling problems.

Considering the same table on a line-by-line basis, the scenario is not better. Among all attacks, only two are confirmed to apply to a single vendor only (i.e., disable feed hold and theft via screenshot). On the other hand, six attacks are confirmed on all vendors.

---

[9] When an attack is reported multiple times is because it consists of variations of the same attack. For example, "change tool geometry" can be leveraged to achieve damage, denial-of-service, or hijacking; this depends on which geometries are changed, the type of machine and the manufacturing process. Vice-versa, distinct attacks can conduct to the same goal. For example, an attacker can take control of the production of an exposed CNC by hijacking a parametric program, by modifying the geometry of a tool to introduce a micro-defect, or by changing the executed program.

Features like the configuration of the geometry of the installed tools, or the modification of the variables used by a parametric program with values supplied via network are automation-facing options, needed when dealing with complex automation and unsupervised process. Although these requirements are nowadays more common in manufacturing, vendors do not seem to take into account unwanted consequences of these features, thus raising concerns about security.

Compromise. The first class of attacks consists of issues that result in a compromise of the CNC machine. While the focus of our research is limited to domain-specific problems, we also conducted a general assessment of the security posture of the controllers under analysis, including the simulators. For this, we used standard vulnerability assessment tools like Nessus with the aid of manual analysis and inspection.

Our experiments confirmed that several CNCs were prone to compromise at different levels including obsolete software or operating systems, weak OEM passwords or service credentials, enabled jumpers that allowed firmware extraction. Considering that our tests were conducted on CNCs ready to be delivered to the end-users, this reveals a general lack of awareness with respect to security.

Damage. This class of attacks consists in damaging either the machine (or part of the machine, such as the tool or the spindle), or the part in production. CNCs are costly machines, with prices ranging from a few thousand to millions of US dollars, so damage is an important issue. Not only is the damage to be considered in terms of breakage of machinery components, and therefore the economic burdens on the end-user, but some interventions to replace the damaged elements also require procurement of complex assemblies, with logistic times usually on the order of weeks or months. Furthermore, the replacement interventions of these components require days of work and a phase of zeroing of the geometries of the machine tool (for example, the setup of the axes), thus introducing, in addition to the monetary cost, an impediment in terms of use of the machine for varying times.

We identified five attacks that could lead to damage. Due to the lack of space, this paper will present two of them.

Feed hold is a functionality that enables an operator to pause the execution of a machine, by stopping the feed axes, for example, to inspect the part in production during a program run. In our experiments, we confirmed that one vendor, Haas, is vulnerable to an attack in which a malicious user can remotely disable the feed hold while being used: an operator pressing the pause button of the machine will not be able to pause the manufacturing. For this vendor, the attack involves abusing the lack of authentication and access control on Ethernet Q to set the global variable 3004 to 7.

Another attack consists of tampering with the geometry of the tools. Each tool used by a CNC needs to be measured in any of its fundamental geometric quantity, depending on the type of machine and manufacturing process. A correct measurement is a must in computing the quotes for working a part within tolerance. In addition to that, any manufacturing process consumes the tool, for example, by reducing the overall geometry of the cutting edge. To address this need, a parameter called *wear* is used as a form of compensation. For example, in the case of a vertical milling machine used to drill holes in a raw part, a negative wear causes the column to crash into the part with damage on the tool or the spindle. Unfortunately, we found that this attack successfully works in all its variations and on all manufacturing controllers, including simulators and real-world installations as we demonstrate in Sect. 5.

Denial-of-Service. Miscreants are often interested in sabotaging the operations of a targeted organization, such as a competitor or a generic victim they can profit from, for example, by demanding a ransom to restore the normal functionalities. With DoS, we mean all attacks aimed at disrupting the manufacturing process, for example, by stopping the machines from operating, or at slowing down the production with the end goal of reducing the efficiency of the industrial process.

We identified six attacks leading to DoS. One of these consists of lowering the load parameter associated with a tool, in order to slow down the production. This attack works because the controller automatically tunes the spindle's speed according to the capacity of the tool installed on the machine.

Another way of causing DoS is triggering alarms so as to block the current execution and request the intervention of the operator. Unfortunately, our evaluation reported that two vendors permit generating software alarms remotely. Although this feature can, to a certain extent, make sense in the development of a program for CNC applications, for example, for a program to trigger an alarm in certain conditions, it is arguable whether it would make sense to offer this option through a remote network call.

We also confirm the possibility to ransom the machines under test by compromising and installing an add-in that locks the HMI (Okuma), or by encrypting the G-code programs exposed via network shares, which were by default unprotected on Haas and Heidenhain.

Hijacking. With hijacking we refer to the possibility for a miscreant to either introduce a micro-defect in the manufacturing process, or to replace the program in execution with one of her choice. In our experiments, we confirmed that all vendors were vulnerable to a change of a tool geometry aimed at introducing a micro-defect. With this, an adversary can take control of the manufacturing process to introduce very small micro-defects that might pass the QA process. These would eventually result in big financial or reputational losses for the victimized manufacturer.

Another option for hijacking the production is to alter the logic of a parametric program. By substituting the values of the memory variables used by a parametric program, an attacker can influence the final outcome. An example of this attack is the production of components "in sizes", in which the difference in geometry is often controlled by the selection (that is, activation) of specific program blocks for manufacturing the size or configuration of the work geometry in a parametric way. The modification of these values leads to the introduction of defects or to the production of wrong sizes compared to what is set by the operator on the HMI. All controllers were affected by this issue.

Finally, on three of the four controllers, we managed to replace the executed program with one of our choice without requiring any operator intervention or notice.

Theft. Theft is a major concern in the manufacturing world. Production includes sensitive information that a manufacturing process produces and that an adversary is interested in monitoring or stealing. In our evaluation, we confirmed that all tested vendors expose such private information to varying degrees. The information we confirmed being exposed within the tested machines includes how many parts are produced, the name of the program associated with each production, the name of the machine, its serial number and related controller version, the active screen or menu on the HMI, the tool number, and part program comments.

Program files constitute a highly sensitive intellectual property because they specify the movements that a machine has to perform to conduct the machining. If an adversary manages to get access to these files, she could reproduce the part on her side or learn all the details behind the manufacturing, as in the case of an adversarial competitor. Theft of program files becomes of even greater concern in consideration that programs developed in G-code are not compiled. In our work, we managed to leak the content of the executed program on three controllers. In all the cases, we performed the attack via network, that is, without the need to bypass any security mechanism like brute-forcing an authentication procedure. In the case of Okuma, the MTConnect service exposes by default the block line currently executed, thus enabling an attacker to poll the daemon to reconstruct the code. For Heidenhain, its DNC interface is by default unauthenticated and a user can therefore remotely dump the executed program (via RPC or LSV2). Similarly, Fanuc exposes such data via Focas.

Finally, the DNC interface of Heidenhain can be abused to take screenshots of the operator's HMI. This enables a miscreant to spy on the manufacturing process, potentially accessing information such as the part program code, the tools list, or the machine configurations in an even more simplified way.

5   Use Cases

In this Section, we provide few a examples among the many attacks that we conducted on our real-world CNC installations, showing how we implemented them and discussing their practical impact[10].

The first experiment consists of abusing the Ethernet Q Commands interface of Haas to conduct three attacks: introducing a micro-defect in the manufacturing process (hijacking), performing a DoS, and damaging a tool. These attacks are possible because Ethernet Q Commands allows for altering the geometry of

[10] An extensive list of use cases are provided in our technical report [4].

Fig. 2. The Haas Super Mini Mill engraving the first trace.

(a) Correct engraving.

(b) Defective engraving.

Fig. 3. Example of hijacking attack.

a tool remotely. As previously mentioned, the controller exposes this interface by default and does not provide authentication nor resource access control.

We conducted these attacks on a Haas Super Mini Mill machine – shown in operation during our experiment in Fig. 2. For this experiment, we developed a program that instructed the machine to engrave four *equal traces* in a part of raw metal. The engraving was supposed to be 5.05 mm deep, as measured in Fig. 3a. The result of the manufacturing cycle is shown in Fig. 4. The part on the left shows the correct execution of the manufacturing, with four traces of the same depth.

At this point, we ran our attacks by altering the wear parameter three consecutive times. First, we set a wear of +0.25 mm on tool number 1 to introduce a micro-defect:

```
$ echo "?E2201 0.25" | nc <IP> 5000
```

Fig. 4. The correct process on the left, and that of our confirmed attacks on the right.

Fig. 5. The 3D-printed plastic tool for our damaging experiment, which crashed against the raw material (left), and a detail thereof (right).

Then, we set the same wear to +5.50 mm, which is more than the original depth of the engraving. Finally, we set the wear to -10 mm.

The result of our attacks is shown in the right part of Fig. 4. This part shows only two engravings instead of four. The first engraving is the reference one and corresponds to the normal execution of the machine. The second engraving has a depth of only 4.80 mm as measured in Fig. 3b, i.e. with an error of 0.25 mm as per attack.

The other two engravings were not made because: In one, the machine operated above the plane of the raw part due to the wear being higher than the depth (5.50 mm > 5.05 mm); in the other, the machine crashed the tool against the raw part because of the negative wear (−10 mm). For this last attack, we printed a plastic tool with a 3D printer, which we voluntarily broke against the raw part during the attack as shown in Fig. 5.

With this single experiment, we demonstrated how an attacker can remotely alter the geometry of tools to conduct attacks with three goals: hijacking the production to insert a micro-defect, making the machine operate above the plane of the material (DoS), and damaging the production's tool or part.

The next experiment shows how to leak the program code running on the machine. Three tested controllers were affected by this issue. In the case of Okuma, the agent reports several useful information related to the manufacturing process like the number of installed tools or the position of the axes. The

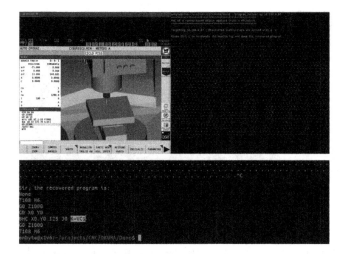

Fig. 6. The dump of the executed program's source code via an unauthenticated and exposed MTConnect agent.

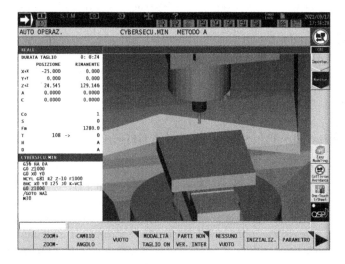

Fig. 7. A parametric program executing two holes as per legitimate operation.

problem lies with the fact that the same agent reports both the name of the executed program and the code block (i.e., the instruction) currently executed on the machine. As result, a miscreant can pool the service to fetch the executed instructions shown in Fig. 6. This is a severe issue because it required nothing more than connecting to the exposed service for conducting the attack. We communicated this issue to Okuma, which promptly acknowledged and fixed it.

One important consequence of being able to dump the executed program is the act of reverse-engineer it, which is fairly easy with G-code. This, leads to

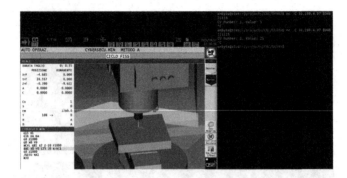

Fig. 8. The same parametric program executing 25 holes after hijacking.

the next use case: parametric program hijacking. As we discussed previously, it is a common practice of developers to use variables to dynamically change the execution flow of a program (as in a sort of conditional IF). In our example, we have a program that is supposed to drill K holes, where K is controlled by the variable VC1, as we highlighted in the instruction block of Fig. 6. In this use case, K holds a value of 2 and the machine drills two holes, as shown in Fig. 7.

At this point, an attacker that understands the program can remotely replace the content of the variable with an arbitrary value (such as 25) in order to hijack the production. This would alter the production to suit the attacker's needs, slowing down the production, or damaging it. Figure 8 shows this example in practice. All tested controller are affected by this issue.

6   Responsible Disclosure and Mitigations

In conducting this research, we wanted to raise awareness in a domain in which security didn't, yet, seem to considered an important driver. With this goal in mind, we underwent an important disclosure process and communicated our findings in a timely and responsible manner with the vendors of the tested controllers. This process was not easy, and required strong commitment on our side in engaging with the right peers and educating them on the importance of the issues that we identified. The large amount of demo material that we collected during our experiments helped in this direction.

Fortunately, all vendors acknowledged our concerns and most of them have addressed, to various degrees, our findings in a reasonable time frame. More importantly, all of them have expressed interest in our research and have decided to improve either their documentation or their communication efforts with the machine manufacturers, with the final goal of offering to the end-users more secure solutions.

Table 4 provides a short summary of this process. CISA's ICS-CERT extended invaluable help and support during our discussion with the vendors, for which we are grateful.

Table 4. A summary of our responsible disclosure process with the vendors.

Vendor	Issues (and CVEs)	Contact Date	Ack Date	Feedback
Haas	Abuse of Ethernet Q Commands (CVE-2022-2474, CVE-2022-2475, CVE-2022-41636). RCE via Java JMX. Firmware extraction via enabled boot jumper	17/11/21 (direct). 13/01/22 (CERT)	20/07/22	Issues acknowledged and public advisory released. The simulator won't be fixed because out of scope
Okuma	RCE via CVE-2010-2729. Abuse of THINC-API. Code leak via MTConnect	19/11/21 (direct)	25/11/21	Issues acknowledged and MTConnect fixed. THINC-API won't be fixed due to performance reasons
Heidenhain	Abuse of DNC (CVE-2022-41648). Weak OEM password. Multiple known vulnerabilities	04/02/22 (direct). 01/03/22 (CERT)	10/05/22	Issues acknowledged and public advisory released
Fanuc	Abuse of Focas.	07/03/22 (direct). 29/03/22 (CERT)	27/04/22	Issues acknowledged and documentation enhanced. Added support for authentication.

We also propose mitigation strategies for both manufacturers and end-users/integrators. With regards to controller manufacturers, we recommend adding support for authentication on all services and enforcing authentication by default. Additionally, we encourage manufacturers to adopt appropriate authorization schemes in the design of their systems, such as privilege separation and access management.

For integrators and end-users, we suggest the following mitigation strategies: Use of context-aware IPS/IDSs that regularly keep up with newer industrial protocols. Correct network segmentation should be implemented to isolate CNC machines from other network assets. Consider modern CNC machines as part of an organization's IT assets and follow the same patch management procedures as any other equipment, such as desktop computers or servers. In our research, we also collaborated with a vendor to add support for proprietary CNC protocols

7   Related Work

While previous work has addressed the security of smart manufacturing technologies, including CNCs to a limited extent, our extensive evaluation of the CNC domain using both controller simulators and real-world machines sets our research apart as the first of its kind.

Quarta et al. [10] conducted a security analysis of an industrial robot. By using a real-world industrial robot, the authors analyzed its architecture and evaluated the associated risks. However, this paper differs from our work in the following ways: firstly, our work focuses on the overall ecosystem of computer numerical controls while this paper focuses on a single robot and its implementation; secondly, our work includes the analysis of CNC machines which differ significantly from industrial robots in terms of design, architecture, and implementation of both software and protocols; thirdly, manufacturers of industrial robots such as ABB, do not typically offer CNC solutions (and vice versa), highlighting the substantial differences between these two types of machine tools.

In a follow-up study, Pogliani et al. [9] explored the security risks associated with bad practices in code development for modern industrial robots. The authors proposed a static-code analysis tool to detect security vulnerabilities in robot code and used it to show that certain implementations of programs found online were effectively vulnerable to different classes of attacks. This work differs in focus from ours. Additionally, the programming languages used in industrial robots (e.g. RAPID and KRL) are quite different from those ones in the CNC domain (G-code, M-code, proprietary macros). Maggi et al. [7] investigated how smart factory floors are exposed to potential security threats in Industry 4.0. They reported security issues at different levels including abusing industrial add-ins or compromising digital twins in software simulators. Their research explored the risks of the industrial ecosystem as a whole, showing that modern smart installations give rise to a larger attack surface, compared with previous generations of industrial facilities. This work touches on the security of the different systems without going vertical on a single category. In addition, the problems identified related with common OS functionalities rather than domain-specific features. Balduzzi et al. [2] looked at industrial gateways used in smart factories to enable communication between modern and legacy devices. The authors reported issues in which translations occurred for example from Modbus TCP to RTU. Niedermaier et al. [8] showed how PLCs can be influenced by packet flooding. The authors conducted an experiment with 16 devices from six vendors, and demonstrated that all except for one device are susceptible to network flooding attacks. Maggi et al. [6] looked at the radio protocols used to remotely control industrial machinery. Their research indicated that multiple vendors were prone to the same class of problems: the ability for a miscreant to arbitrarily generate fake radio messages and sabotage the operation of industrial plants. Similar problems were reported by Balduzzi et al. [3] who conducted a security analysis of a radio protocol standard used in the maritime industry for monitoring and tracking logistics and passenger ships.

In a work closer to ours, Chen et al. [5] discussed the hypothetical risks associated with CNC machines, reporting issues related to a CNC's terminal. The authors proposed mitigation strategies like the adoption of cryptographic schemes for data protection, or industrial gateways for proper network segmentation and access control. Similarly, Tu et al. [11] proposed a trusted security framework for CNC machines. Although these works sit in the same domain of research as ours, they provide different research methodologies and contributions. Our work is closer to the real-world implementations of CNC machines, in having conducted an empirical evaluation of the security boundaries of the technologies put in place by controller manufacturers according to the needs dictated by Industry 4.0.

8   Conclusions

Our research explored the risks associated with the adoption of Industry 4.0 in CNC machines. These machines underwent a shift from standalone systems to network-enabled ones that resemble full-fledged machines more closely than they do mechanical devices. As a result, end-users are left dealing with sophisticated systems that, if not correctly configured or poorly designed, might open the door to abuse.

In our research, we explored technologies specific to the CNC domain and conducted an extensive security evaluation. We implemented PoC attacks on real-world installations, demonstrating that our concerns have practical implications, and identified important issues that are common among all controllers under test.

In addition to publishing our findings in this research paper, we also created demo material to educate the community about the security risks in the CNC domain. Our responsible disclosure process prompted interest from the affected manufacturers, who acknowledged our findings. Our aim is to raise awareness in a field that we believe will gain more attention in the future.

References

1. Interchangeable variable block data format for positioning, contouring, and contouring/positioning numerically controlled machines. Electronic Industries Association (1979)
2. Balduzzi, M., Bongiorni, L., Flores, R., Lin, P., Perine, C., Vosseler, R.: Lost in translation: when industrial protocol translation goes wrong. Trend Micro (2020). https://www.madlab.it/papers/wp-lost-in-translation-when-industrial-protocol-translation-goes-wrong.pdf
3. Balduzzi, M., Pasta, A., Wilhoit, K.: A security evaluation of AIS automated identification system. In: Proceedings of the 30th Annual Computer Security Applications Conference (2014)
4. Balduzzi, M., Sortino, F., Castello, F., Pierguidi, L.: The security risks faced by CNC machines in industry 4.0. Trend Micro (2022). https://www.madlab.it/papers/cnc.pdf
5. Chen, X., Wang, Z., Yang, S.: Research on information security protection of industrial internet oriented CNC system. In: 2022 IEEE 6th Information Technology and Mechatronics Engineering Conference (ITOEC) (2022)
6. Maggi, F., et al.: A security evaluation of industrial radio remote controllers. In: Perdisci, R., Maurice, C., Giacinto, G., Almgren, M. (eds.) DIMVA 2019. LNCS, vol. 11543, pp. 133–153. Springer, Cham (2019). https://doi.org/10.1007/978-3-030-22038-9_7
7. Maggi, F., et al.: Smart factory security: A case study on a modular smart manufacturing system. Procedia Comput. Sci. **180**, 666–675 (2021)
8. Niedermaier, M., et al.: You snooze, you lose: Measuring plc cycle times under attacks. In: WOOT@ USENIX Security Symposium (2018)
9. Pogliani, M., Maggi, F., Balduzzi, M., Quarta, D., Zanero, S.: Detecting insecure code patterns in industrial robot programs. In: Proceedings of the 15th ACM Asia Conference on Computer and Communications Security, pp. 759–771 (2020)

10. Quarta, D., Pogliani, M., Polino, M., Maggi, F., Zanchettin, A.M., Zanero, S.: An experimental security analysis of an industrial robot controller. In: 2017 IEEE Symposium on Security and Privacy (SP), pp. 268–286. IEEE (2017)
11. Tu, S., Liu, G., Lin, Q., Lin, L., Sun, Z.: Security framework based on trusted computing for industrial control systems of CNC machines. Int. J. Performability Eng. **13**, 1336–1346 (2017)

Security Issues When Dealing with Users

A Deep Dive into the VirusTotal File Feed

Kevin van Liebergen[1]([✉]), Juan Caballero[1], Platon Kotzias[2], and Chris Gates[2]

[1] IMDEA Software Institute, Madrid, Spain
{kevin.liebergen,juan.caballero}@imdea.org
[2] Norton Research Group, Tempe, USA
{platon.kotzias,chris.gates}@nortonlifelock.com

Abstract. Online scanners analyze user-submitted files with a large number of security tools and provide access to the analysis results. As the most popular online scanner, VirusTotal (VT) is often used for determining if samples are malicious, labeling samples with their family, hunting for new threats, and collecting malware samples. We analyze 328M VT reports for 235M samples collected for one year through the VT file feed. We use the reports to characterize the VT file feed in depth and compare it with the telemetry of an AV vendor. We answer questions such as How diverse is the feed? How fresh are the samples it provides? What fraction of samples can be labeled on first sight? How different are the malware families in the feed and the AV telemetry?

1 Introduction

Online scanners analyze artifacts (i.e., files, URLs, domains, IPs) submitted by users using a large number of security tools, and provide access to the analysis results through free and commercial APIs. The most popular online scanner is VirusTotal [42] (VT), which is widely used by security analysts, and acts as a de-facto central sharing service for the security community. Detection labels in VT reports are routinely used for determining if an artifact is malicious by either applying a threshold on their count (e.g., [27,29,44]) or feeding them to machine-learning models [34,39], as well as for identifying the family of malicious files [16,35,36]. Prior work has shown that VirusTotal can be used to identify new malware before it is released, since malware developers often leverage VT during development to check if their samples are detected and, if so, revise them until they become *fully undetected* (FUD) [13,14,43]. VirusTotal is also commonly used as a source for collecting malware samples [3,12,14,25,26].

Amongst its commercial services, VT offers feeds, i.e., streams of analysis reports for all submissions of a type [1]. VT offers separate feeds for files, URLs, and domains. In this work, we perform what we believe is the first characterization of the VT file feed (or simply the feed). The VT file feed includes reports for new files (i.e., first submission to VT), resubmissions of previously submitted files, and re-scans requested by users. Each *report* in the VT file feed contains detailed information about the analysis of a sample (i.e., file). The report contains, among others, file metadata (e.g., hashes, size), certificate metadata for

© The Author(s), under exclusive license to Springer Nature Switzerland AG 2023
D. Gruss et al. (Eds.): DIMVA 2023, LNCS 13959, pp. 155–176, 2023.
https://doi.org/10.1007/978-3-031-35504-2_8

signed samples (e.g., thumbprint, subject), VT specific data (e.g., time of first submission to VT, submission filenames), and the list of detection labels assigned by up to 70 antivirus (AV) engines used to scan the file. The VT file feed service also allows unlimited downloads of the samples submitted in the last seven days.

We collect reports from the VT file feed for one year, from December 21st, 2020 to December 20th, 2021. During the first 11 months we collect reports where the sample is detected by at least one AV engine, while in the last month we collect all feed reports, regardless of the number of detections. Overall, we collect 328M reports for 235M samples. We analyze the collected reports to characterize the VT file feed as a source for collecting malicious samples and for identifying new threats. Samples from the feed can be used for building labeled malware datasets such as those required by machine learning (ML) based malware detection (e.g., [4,15,17,32,37]) and family classification (e.g., [15,33]). We investigate fundamental questions for such use including How diverse is the feed? Does it allow building malware datasets for different filetypes? How fresh are the samples it provides? What is the distribution of malware families it sees? The feed can also be a source for malware triage and malware hunting approaches (e.g., [10,18]). For this use, we investigate what fraction of the feed samples are variants of known malware families that analysts may not need to investigate. In particular, we measure what fraction of the samples in the VT file feed can be detected as malicious on first sight, what fraction can be labeled with a family on first sight, and what fraction of malicious samples are originally fully undetected but later become detected by multiple AV engines. We complement our characterization of the VT file feed with a comparison with telemetry data collected in a privacy-sensitive manner from tens of millions of Windows devices of clients of a large antivirus vendor. The comparison allows us to investigate how different are the views of the malware landscape observed by both datasets and which dataset observes samples faster.

To improve family labeling, we have more than doubled the size of the AvClass [36] taxonomy and tagging rules. We have contributed our updates to the AvClass repository and they have been integrated into AvClass 2.8.0. The following are some of the most significant insights we gain:

- The VT file feed is a great source for malicious samples with a much higher maliciousness ratio than the AV telemetry. Still, the VT file feed is not a malware feed since half of its volume is for benign samples. Thus, it can be used to build both malicious and benign file datasets for supervised ML approaches.
- The feed is diverse with a wealth of filetypes and 4.9K families with at least 100 samples. However, the diversity is largely due to Windows and Android families.
- The feed is fresh: it receives an average of 732K new malicious samples each day and malicious samples appear a median of 4.4 h after they are seen in user devices. 39% of new malicious samples appear in the VT file feed earlier than in the AV telemetry, allowing AV engines to leverage the VT file feed to build detections for samples before they affect their customers.

Table 1. Dataset collected from VT file feed between 2020/12/21 and 2021/12/20.

Data	All	peexe	apk	other
Reports	328.3M	220.3M	15.9M	92.0M
Samples	235.7M	155.5M	8.2M	72.0M

- On first sight, 62% of the samples can be labeled as variants of known families, and thus could be ignored when hunting for new threats.
- We identify 600K originally FUD samples. These samples have no detections on first sight, but are later detected by multiple AV engines.
- The AV telemetry and VT file feed observe largely disjoint sets of malicious samples with minimal overlap (1.2%–1.8%).
- The most popular families in the VT file feed by number of samples widely differ from the families affecting most devices in the AV telemetry.

2   Datasets

We use two datasets in this work. We collect reports of files that appear in the VT file feed for one year. We also examine the Windows telemetry of an AV vendor over the same time period, which contains the metadata (e.g., file hash, file type) of the files present in tens of millions of Windows devices that opted-in to the data collection. Both datasets include benign and malicious files of different file types.

VT File Feed. The VT file feed contains analysis reports for files submitted to VT, regardless of the file type and platform (e.g., Windows executables, Android APKs, Linux ELF executables, PDF and Microsoft Office documents). Other artifacts submitted to VT like URLs, domains, and IPs have their own separate feeds that we do not analyze. The VT file feed includes reports for new files (i.e., first submission to VT), resubmissions of previously submitted files, and user-requested re-scans of previously submitted files. Throughout the paper we use *sample* and *file* indistinctly. Multiple reports may appear in the feed for the same file. In general, we focus on the last report we collected for each sample because it should provide the most up-to-date information (e.g., updated AV labels). However, when interested in what happened to a sample when first submitted to VT (e.g., whether it was detected or labeled), we examine instead its first report.

We collect reports from the feed every minute. To keep the storage manageable, we do not download the samples from the feed, only the reports. In the first 11 months, we only collected reports where at least one AV engine detected the file as malicious, which (as later shown) roughly corresponds to half of all reports in the feed. On November 19th, 2021, we started collecting all reports in the feed regardless of the number of detections, i.e., including reports with zero detections. Overall, as summarized in Table 1, over one year between December 21st, 2020 and December 20th, 2021, we collected 328M reports for 235M samples (by unique file SHA256).

Table 2. Features used.

Feature	Scope	Type	peexe	apk
cert_issuer	sample	string	✓	✓
cert_subject	sample	string	✓	✓
cert_thumbprint	sample	cryptohash	✓	✓
cert_valid_from	sample	timestamp	✓	✓
cert_valid_to	sample	timestamp	✓	✓
exiftool_filetype	sample	string	✓	✓
fseen_date	sample	timestamp	✓	✓
md5	sample	cryptohash	✓	✓
package_name	sample	string	✗	✓
sha1	sample	cryptohash	✓	✓
sha256	sample	cryptohash	✓	✓
trid_filetype	sample	string	✓	✓
detection_labels	scan	string list	✓	✓
scan_date	scan	timestamp	✓	✓
sig_verification_res	scan	string	✓	✗
vt_meaningful_name	scan	string	✓	✓
vt_score	scan	integer	✓	✓
avc_family	derived	string	✓	✓
avc_tags	derived	string list	✓	✓
avc_is_pup	derived	bool	✓	✓
filetype	derived	string	✓	✓

Telemetry. The telemetry comprises metadata of files present in tens of millions of real Windows devices in use by customers of an AV engine. It does not contain the samples, only their metadata. The customers opted-in to sharing their data and the devices are anonymized to preserve customer privacy. The AV engine queries a central service with file hashes observed on the device to obtain file reputation information. Each query for a file hash sent by a device is an *event*. An event comprises a timestamp, the anonymous identifier of the device, a file hash, a filename, and the signer key if the file is signed (i.e., the SHA256 of the public key in the file's certificate). The telemetry contains events for both benign and malicious files present on the devices. Those files may be of different types including Windows PE executables (e.g., .exe, .dll, .sys, .ocx), PDF documents, and Microsoft office files. We also obtain information from the AV vendor on the subset of telemetry files for which the AV engine threw an alert, i.e., the detected samples. We examine telemetry events over the same one year period we monitored the VT file feed.

3 Features

Since we do not download the samples, we need to restrict our analysis to features available in the reports, or that can be derived from the reports. We focus on a selected set of 21 features: 17 from the VT reports and 4 derived from those (e.g., filetype and malware family). Features are summarized in Table 2. We define three scopes for a feature: sample, scan, and derived. Sample features should have the same value across all scans of a sample. On the other hand, scan features may differ across scans of the same sample, i.e., they evolve over time. For example, the hash of the certificate of a signed sample (*cert_thumbprint*) should always be the same. But, whether the signature of a signed sample validates (*sig_verification_res*) can change across scans, e.g., if the certificate expires or is revoked. Features may be extracted only for a subset of filetypes, e.g., be specific to Windows PE executables or Android APKs, and may be null for some samples (e.g., certificate features are not available for unsigned Windows executables). We detail the VT report features in Sect. 3.1 and the derived features in Sect. 3.2.

3.1 VT Report Features

Of the 17 features from the VT report, 3 are cryptographic hashes over the whole file used to identify the sample (*sha256*, *sha1*, *md5*), 5 are related to code signing, 2 capture the file type, another 2 capture the program name, and 5 are specific to the scan. The code signing features are available for a variety of file types including Android APKs, iOS apps, signed Windows executables, and signed Windows MSI installers.

Timestamps. We obtain four timestamps from a VT report. The *scan_date* when the sample is analyzed, which is always within our collection period. The VT first seen date (*fseen_date*) when the sample was first submitted to VT. For signed samples, we also obtain the certificate's validity period defined by the *cert_valid_from* and *cert_valid_to* dates.

AV Scans. VT scans each submitted sample with a large number of AV engines. We extract the number of engines that detected the sample (i.e., gave it a non-NULL label) (*vt_score*) and the list of *detection_labels*. The labels are used to derive three classification features, as detailed in Sect. 3.2.

Program Names. We use two features that capture the program a sample corresponds to. The *package_name* is the package identifier for Android apps and *vt_meaningful_name* is the most meaningful filename VT selects for a sample (e.g., among all filenames of the sample when submitted to VT).

3.2 Derived Features

Filetype. Determining the filetype of the sample in a report is not straight-forward because VT reports do not have a single field for it. Instead, there are multiple fields that provide, possibly contradictory, filetype information. We derive a unique *filetype* feature for each report by performing a majority voting on three fields: *trid_file_type*, *vt_tags*, and *vt_meaningful_name*. *trid_file_type* captures the filetype identified by the TrID tool [31], which has very fine-grained granularity (e.g., over 90 *peexe* subtypes). We build a mapping from TrID file-types to coarser-grained filetypes such as grouping all Windows PE files (e.g., EXE, DLL, OCX, CPL) under *peexe* and all Word files (DOC, DOCX) under *doc*. *vt_tags* provides a list of tags assigned by VT to enable searching for samples across different dimensions. Some of the tags such as *apk*, *peexe*, and *elf* provide filetype information. When *vt_meaningful_name* is available, we extract the extension from the filename and map the extension to a filetype.

AVClass Features. We feed the *detection_labels* to the AVCLASS malware labeling tool [36]. AVCLASS outputs a list of tags (*avc_tags*) for the sample that include its category, behaviors, file properties, and the most likely family (*avc_family*). It also provides whether the sample is considered potentially unwanted or malware (*avc_is_pup*). AVCLASS uses a taxonomy to identify non-family tokens that may appear in the AV labels such as malware classes (e.g., *CLASS:virus*), behaviors (e.g., *BEH:ddos*), file properties (e.g., *FILE:packed:asprotect*), and generic tokens (e.g., *GEN:malicious*). It also uses tagging rules to identify aliases between families (e.g., *zeus* being an alias to *zbot*). In this work we apply AVCLASS to 328M VT reports, eight times more than the largest to date work [36]. Thus, our AVCLASS results include a wealth of new tags, including new aliases and non-family tokens. We have used the AVCLASS update module and extensive manual validation to identify new tagging rules that capture previously unknown aliases, as well as new taxonomy entries for tokens appearing in over 100 samples. This process has resulted in more than doubling the AVCLASS taxonomy and tagging rules. We have contributed our updates to the AVCLASS repository.

4   Feed Analysis

This section characterizes the VT file feed, answering the following questions: (1) How large is the VT file feed? (2) How fresh are samples in the feed? (3) How diverse is the feed in terms of filetypes? (4) What fraction of samples are signed? (5) What fraction of samples can be detected as malicious on first scan? (6) What fraction of malicious samples are fully undetected on first scan? (7) How diverse is the feed in terms of families? (8) What fraction of samples can be labeled on first sight?

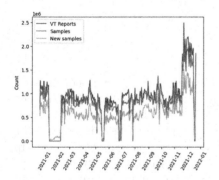

Fig. 1. Number of daily VT reports and samples collected.

Fig. 2. Number of samples first seen by VT on each month. y-axis is in logarithmic scale.

Table 3. Daily statistics when collecting all reports (from 2021/11/21 to 2021/12/20).

	Mean	Median	Stdev	Max
Reports	1,786,565	1,879,952	482,286	2,492,454
Samples	1,586,750	1,680,520	424,590	2,223,638
New samples	1,092,640	1,120,242	299,645	1,504,174

Volume. Figure 1 shows for each day in the collection period, the number of reports in the feed, the number of unique samples in the daily reports, and the number of samples first seen by VT on that day. The figure shows a few gaps when the collection infrastructure was not working, the longest taking place between January 11th and February 7th. The volume of reports and hashes significantly increases once we started collecting samples with no detections. We compute the daily statistics, excluding days in the collection gaps, split into two periods: before November 21st, 2021 when we were collecting only reports with at least one detection, and after that date when we were collecting all reports. We say that a sample is *new* only on the first day that it is submitted to VT. Table 3 shows the daily stats when collecting all reports: the average number of daily reports is nearly 1.8M, the average number of samples nearly 1.6M, and the average number of new samples nearly 1.1M. When only collecting reports with at least one detection the daily averages were 913K reports, 823K samples, and 580K new samples. Thus, approximately half of the reports (51%), samples (51%), and new samples (53%) in the feed are for undetected samples.

> **Takeaway 1**
> At the end of 2021, the VT file feed had daily averages of 1.8M reports, 1.6M samples, and 1.1M new samples. The VT file feed is a file feed rather than a malware feed. Half of its volume in terms of reports, samples, and new samples is for undetected samples.

Freshness. The same sample may appear in the VT file feed multiple times, e.g., because different users submit it at different times. On average, 69% of the files observed in one day are new (i.e., previously unknown to VT) and 31% correspond to re-submissions or re-scans of already known files. Over the one year analysis period, 89% (209M) of the samples had a VT first seen date later than our collection start date. This ratio increases over time as every day the influx of new samples (69%) is larger than that of already seen samples (31%).

The previously seen samples that re-appear in the feed may be fairly recent or really old. The VT first seen date provides a lower bound for a sample's lifetime, i.e., the sample could be older if it took some time for it to be submitted to VT. The oldest sample observed in our collection period was first seen by VT on May 22nd, 2006. Figure 2 shows the number of samples (in logarithmic scale) whose VT first seen date is on each month, capturing how old are the samples already known to VT. The shape of the figure captures the volume increase in samples

Table 4. Top 20 filetypes for all samples observed. *peexe* includes all Windows PE files (EXE, DLL, CPL, OCX, ...) *doc* and *xls* include also *docx* and *xlsx*, respectively. *NULL* corresponds to samples for which a filetype could not be determined.

#	Filetype	Samples	Perc		#	Filetype	Samples	Perc
1	peexe	155,526,594	65.97%		12	elf	942,148	0.40%
2	javascript	21,048,404	8.93%		13	rar	516,514	0.22%
3	html	12,540,571	5.32%		14	jar	448,324	0.19%
4	pdf	11,346,815	4.81%		15	doc	429,794	0.18%
5	apk	7,992,206	3.40%		16	xls	428,057	0.18%
6	text	5,149,050	2.18%		17	macho	409,399	0.17%
7	NULL	4,128,183	1.75%		18	php	352,143	0.15%
8	zip	3,934,987	1.67%		19	xml	335,962	0.14%
9	dex	3,015,650	1.28%		20	powershell	321,178	0.14%
10	gzip	2,926,739	1.24%			Other	1,233,754	0.52%
11	lnk	2,718,635	1.15%			ALL	235,745,107	100.0%

submitted to VT over time until 2019, followed by a decrease in 2019–2021. The reduction could be due to some vendors reducing their sharing from 2019.

Takeaway 2

On average, 69% of the samples observed in one day are new, i.e., previously unknown to VT, and the feed provides over a million new samples each day. Thus, the VT file feed is a great source of fresh samples.

Filetypes. Table 4 shows the top 20 filetypes for all samples observed. The feed is dominated by Windows PE files (EXE, DLL, OCX, CPL, ...) that correspond to 66% of the samples. Far behind are other filetypes like JavaScript (8.9%), HTML (5.3%), PDF (4.8%), and Android applications (3.4%). The top 5 filetypes cover 88.4% of all samples. We could not obtain a filetype for 1.7% of samples as they had no TrID information, no VT filetype-related tags, were not signed, and had no most meaningful filename with extension. This highlights the lack of a unified filetype field and the limitation of the tools VT uses for filetype determination.

Ugarte-Pedrero et al. [40] reported that 51% of an AV feed were PE executables. The larger VT file feed ratio may be due to users contributing more frequently PE executables to VT, avoiding other filetypes like HTML or text files that may contain more private data.

Takeaway 3

Two thirds of feed samples are Windows PE files, but the feed is a good source of samples for a large variety of filetypes. The feed lacks a unified filetype field and filetype identification is challenging for a significant number of samples.

Code Signing. VT extracts code signatures from multiple filetypes. The collected reports contain 13.3M samples (5.6% of all samples) for which VT extracted code signing certificates. Of the signed samples, 55.9% are Android APKs, 43.4% are Windows PE files, and 0.7% are other filetypes such as Microsoft Installers (.msi) and patches (.msp), Mach-O executables, iOS applications, Apple image files (.dmg), and some archive formats (e.g., .zip, .cab). PDF is one popular filetype for which VT does not currently extract signatures. 91.3% of all *apk* samples, 3.7% *peexe*, 31.4% *msi*, and 7.6% *macho* are signed. APKs have to be signed in order for the Android OS to install them in a device. The 8.7% of unsigned APKs is due to apps under development being uploaded to VT, possibly to check if any AV engine detects them or as part of continuous delivery pipelines.

> **Takeaway 4**
>
> VT supports the extraction of code signatures for a variety of filetypes, but only a small fraction (5.6%) of all feed samples, and 3.7% of the *peexe* samples, have a code signing signature.

4.1   AV Detections

A common approach for detecting malicious samples is to apply a threshold on the number of detections in a VT report [44]. We use this approach to quantify the percentage of malicious samples in the feed. We focus on the last month when collecting all feed reports. Figure 3 shows the distribution of the number of AV detections for all reports collected starting 2021/11/21. The figure shows that 51% of the reports in the last month have no detections and 7% have one detection. But, there are 9.6M samples with at least 40 detections.

We also examine the number of detections the first time a sample is submitted to VT. Figure 4 shows the complementary CDF of VT scores for the first report of each new sample since 2021/11/21. The figure captures the fraction of malicious

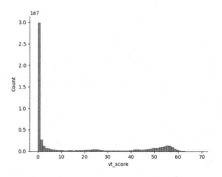

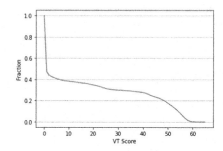

Fig. 3. Number of detections distribution for all reports since 2021/11/19.

Fig. 4. Reverse ECDF for the first report of each new sample since 2021/11/19.

samples in the feed depending on the selected detection threshold. 53% of the samples have zero detections on their first observation. This percentage includes truly benign programs as well as malicious samples that go fully undetected. If we set the detection threshold on at least one detection, 47% of the samples would be considered malicious. If the threshold is set higher to minimize false positives, that reduces the fraction of malicious samples, e.g., 41% if we set it to at least four detections as done in several related works [19–21].

> **Takeaway 5**
>
> On first sight, 41% of samples are detected as malicious by at least 4 AV engines, and 47% by at least one AV engine. These malicious samples share traits with previously seen malware (i.e., match existing signatures).

Originally FUD Malware. It is possible that a malicious sample is fully unde-tected when first submitted to VT, but a later report classifies it as malicious. To detect originally FUD samples, we measure the number of samples that sat-isfy three conditions: (1) they are first observed by VT during our collection period; (2) their last report has at least 4 detections; and (3) their first report had zero detections *or* their VT first seen date is not in a data collection gap and is before their first observation. The last condition is a disjunction to address that we only collected reports with zero detections in the last month. During the first 11 months we can know if a sample had zero detections in their first scan because their VT first seen date is in our collection period and happens before the earliest scan date collected for the sample. The exception are samples first seen during a collection gap, for which a delayed scan date does not necessarily imply zero detections on the first scan.

We identify 637K samples satisfying those conditions. However, the time difference between the first seen date and the first report with at least four detections, indicates that 37K samples change from zero to at least four detec-tions within 5 min of their first VT observation. We exclude those 37K samples as we observe that the distribution stabilizes afterwards (i.e., after 15 min only an extra 1K samples flip classification).

Thus, we identify 600K originally FUD samples that had no detections on their first scan, but were later considered malicious by at least 4 AV engines. Increasing the detection threshold would decrease the percentage, but the detec-tion rate of a malicious sample tends to increase over time and for 82% of samples we only have one report. Thus, we believe our FUD rate estimation is conserva-tive. The median time to flip classification is 7 days, (mean of 23.8 days) with 12% of the samples flipping classification in less than one day.

Of the 600K originally FUD samples, 60% are *peexe*, followed by 11% *pdf*, and 8% *javascript*. PDFs are more than twice as likely to be FUD than expected since they comprise only 4.8% of all feed samples. Malicious PDFs typically contain exploits and are used in spearphishing attacks. These numbers point to malicious PDFs being harder to detect.

Takeaway 6

Over the one year analyzed, we identify 600K samples that are originally FUD, i.e., they have zero detections on the first VT observation,
but later are considered malicious by at least 4 engines. PDF documents are more likely to be FUD than other filetypes.

4.2   Family Labeling

We obtain a sample's family by feeding to AVCLASS the last report of each
sample in our dataset, which should have the most up-to-date labels. AVCLASS
labels 151.7M (64.3%) of the samples with 74,360 distinct family names. However, many families output by AVCLASS are rare. In particular, 41.4K (55.8%)
of all families have only one sample, 14K (19.5%) have at least 10 samples,
4.9K (6.7%) have at least 100 samples, 1.5K (2.1%) have at least 1K samples,
526 (0.7%) have at least 10K samples, 147 families (0.2%) have at least 100K
samples, and only 32 families (0.04%) have at least 1M samples.

Despite more than half of the families having only one sample, the fact that
there are 4.9K families with more than 100 samples shows that the feed is diverse
and is not dominated by a few highly polymorphic families (e.g., file infectors).
However, the diversity is largely due to Windows families. By filetype, the number of families with more than 100 samples is led by *peexe* with 3.8K families,
followed far behind by *apk* (447), *html* (129), *javascript* (116), *doc* (53), *macho*
(52), *xls* (47), *elf* (37), and *pdf* (15). Thus, by monitoring the feed it is possible
to build datasets with a large number of families for Windows and Android malware. But, for other filetypes like *macho* and *elf*, even after collecting for a year,
we could only obtain 52 and 37 families with at least 100 samples, respectively.

AVCLASS outputs as family the top-ranked tag that is either a family in the
taxonomy or unknown (i.e., not in the taxonomy). Of the 74,360 families output
by AVCLASS, 2,391 (3.2%) are in the updated taxonomy, which contains a total
of 2,451 families (i.e., 97.5% of taxonomy families appear in one year of feed
reports). However, the families in the updated taxonomy contribute 90.6% of
the labeled samples, only 9.4% of the samples are labeled with unknown families. This indicates that the most popular families are in the updated taxonomy,
which is expected as it is common for analysts like us to add the most popular
previously unknown families to the taxonomy. In fact, of the families with at
least 1M samples, only 3% are unknown, increasing to 15% for families with
100K samples, 43% for those with 1K samples, and 85% for those with 10 samples. Unknown families can be due to two main reasons. One are tags that it
is unclear if they are a family name or another category such as a behavior
or a file property (e.g., *lnkrun*, *refresh*). The other are tags that correspond to
random-looking signature identifiers or family variants (e.g., *aapw*, *dqan*). We
manually examine the top 1K families and identify that 89% of the unknown
families correspond to the first case and 11% to the latter. We repeat this check
on 200 randomly sampled unknown families with only one sample and the result
is the opposite: 11% corresponding to the first case and 89% to the latter. Thus,

Table 5. Peexe top 10 families.

Family	Class	Samples
FAM:berbew	backdoor	19,371,273
FAM:dinwod	downloader	9,398,314
FAM:virlock	virus	7,921,534
FAM:pajetbin	worm	7,164,373
FAM:sivis	virus	6,222,693
FAM:lamer	virus	4,074,441
FAM:salgorea	downloader	3,737,865
FAM:vobfus	worm	3,415,996
FAM:drolnux	worm	2,858,975
FAM:griptolo	worm	2,407,104

Table 6. Apk top 10 families.

Family	Class	Samples
FAM:smsreg	pup	616,406
FAM:ewind	pup:adware	430,531
FAM:hiddad	pup:adware	219,577
FAM:fakeadblocker	pup:adware	82,715
FAM:airpush	pup:adware	80,704
FAM:revmob	pup:adware	78,495
FAM:dowgin	pup:adware	68,522
FAM:dnotua	pup	65,330
FAM:kuguo	pup:adware	63,262
FAM:mobidash	pup:adware	40,016

Table 7. Elf top 10 families.

Family	Class	Samples
FAM:xorddos	ddos	287,631
FAM:mirai	backoor	163,525
FAM:gafgyt	backoor	59,348
FAM:tsunami	backoor	3,381
FAM:hajime	downloader	2,499
FAM:mozi	backdoor	1,996
FAM:setag	backdoor	1,454
FAM:dofloo	backdoor	890
FAM:fakecop	pup	805
FAM:ladvix	virus	580

Table 8. Mach-O top 10 families.

Family	Class	Samples
FAM:flashback	downloader	33,087
FAM:mackontrol	backdoor	15,459
FAM:mackeeper	pup	15,017
FAM:evilquest	ransomware	7,070
FAM:cimpli	pup:adware	5,444
FAM:gt32supportgeeks	pup	3,453
FAM:genieo	pup:adware	3,339
FAM:bundlore	pup:adware	3,142
FAM:installcore	pup:adware	1,543
UNK:fplayer	pup:adware	905

Table 9. Macros (doc & xls) top 10 families.

Family	Class	Samples
FAM:emotet	infosteal	26,430
UNK:sneaky	downloader	23,521
FAM:qbot	downloader	22,416
FAM:squirrelwaffle	downloader	18,230
FAM:valyria	downloader	16,256
FAM:sagent	downloader	13,298
FAM:zloader	downloader	12,371
FAM:sload	downloader	10,923
UNK:encdoc	downloader	5,703
FAM:thus	virus	4,917

Table 10. Javascript top 10 families.

Family	Class	Samples
FAM:faceliker	clicker	2,288,894
FAM:facelike	–	952,180
FAM:coinhive	miner	766,087
FAM:cryxos	–	744,894
FAM:smsreg	pup	415,669
UNK:gnaeus	–	400,570
FAM:fakejquery	downloader	330,792
UNK:hidelink	–	210,306
UNK:agentwdcr	–	87,101
FAM:inor	downloader	83,694

for less prevalent families AvClass may output a name that corresponds to a signature identifier or variant. While those random-looking names are not very descriptive for analysts, they are still valid cluster identifiers, i.e., samples with the same name should belong to the same family. Based on the above, we estimate that over the whole year a total of 33.8K (41.4K * 0.11 + 32.9K * 0.89) families of all filetypes have been observed in the feed.

We also obtain the family using the first report for samples first seen during our monitoring period. AvClass is able to label on first sight 62.3% of samples,

Table 11. Html top 10 families.

Family	Class	Samples
UNK:refresh	–	882,026
FAM:cryxos	–	363,821
FAM:faceliker	clicker	312,563
FAM:smsreg	pup	201,253
UNK:redir	–	200,926
FAM:coinhive	miner	152,968
UNK:generickdz	–	121,975
UNK:pushnotif	–	120,085
FAM:ramnit	virus	80,044
UNK:fklr	rogueware	79,353

Table 12. Pdf top 10 families.

Family	Class	Samples
UNK:fakeauthent	phishing	194,963
UNK:minerva	phishing	15,527
FAM:pdfka	exploit	13,618
UNK:pidief	exploit	6,319
FAM:alien	downloader	6,137
UNK:gorilla	phishing	4,749
UNK:talu	phishing	2,379
UNK:gerphish	phishing	1,558
UNK:urlmal	phishing	1,469
FAM:rozena	backdoor	839

slightly less than the 64.3% using the last collected report. The fact that 62% of samples can be attributed on first sight to a family indicates they correspond to variants of known families with accurate signatures. This result shows that AvCLASS can be used during triage as a filter to remove 62% of samples from well-detected families so that analysts can focus on the 38% unlabeled samples.

Prior work has applied AvCLASS to *peexe*, *apk*, and *elf* files (e.g., [12,36]). However, AvCLASS can be applied on AV labels regardless of platform or filetype. Tables 5, 6, 7 and 8 show the top 10 families for the four executable filetypes. The largest families overall are for Windows led by *berbew* with 19.4M samples, followed by *dinwod* (9.4M), and *virlock* (7.9M). We use AvCLASS to output a relations file on the whole feed. We identify a family's class checking the strongest CLASS relation for each family with a strength of at least 0.2. The top 10 *peexe* families are dominated by 4 worm and 3 virus families due to their high polymorphism. However, as already discussed, overall the feed is not dominated by file infectors and worms. For Android, the top 10 families are all PUP and 8 of them are adware. The top Linux families are dominated by backdoors including *mirai* derivatives (*gafgyt*, *hajime*, *mozi*). For macOS, seven top families are PUP and five of those adware. Table 9 shows the top 10 families for Microsoft Office macros including both Word and Excel files. Malicious macros are dominated by downloaders. Tables 10, 11 to 12 show the top families for three other popular filetypes (JavaScript, HTML, PDF) for which we observe that top families output by AvCLASS contain many unknown tokens that may correspond instead to other categories (e.g., *redir* may indicate injections that redirect the user). We also observe overlaps between JavaScript and HTML families (e.g., *cryxos*, *facelike*) and that for 9/30 families we cannot identify a class. We conclude that for these three filetypes the concept of a family is not as well defined and that AV labels for these filetypes capture instead behaviors such as phishing, injections, and exploitation.

Table 13. Top 10 families (>10K samples) sorted by ratio of originally FUD samples.

Family	Class	Type	FUD Samp.	Ratio
pcacceleratepro	pup	peexe	1,749	9.5%
sagent	down.	macro	2,141	9.3%
dstudio	down.	peexe	1,255	6.2%
pasnaino	down.	peexe	613	5.9%
opensupdater	pup	peexe	2,051	4.8%
mobtes	down.	apk	967	4.6%
hesv	pup	peexe	849	4.4%
asacub	infosteal	apk	833	4.1%
agentino	down.	peexe	649	4.0%
fakecop	pup	apk	672	3.6%

Table 14. Top 10 families for feed samples in the telemetry ranked by number of infected devices.

Family	Class	Dev.	Samp.
winactivator	pup	2.0M	10,871
utorrent	pup	1.6M	1,366
installcore	pup	1.5M	46,758
webcompanion	pup	1.4M	2,569
dotsetupio	pup	1.1M	198
iobit	pup	898K	4,321
opensupdater	pup	692K	14,918
opencandy	pup	579K	9,346
offercore	pup	555K	363
driverreviver	pup	545K	615

Takeaway 7

The feed is diverse. Over one year, 33K families are observed with 4.9K families having at least 100 samples. However, the diversity is largely due to *peexe* and *apk* families. For those two filetypes, the feed is a good source to build datasets for large-scale family classification. AvClass labels 62% of samples on first sight. Thus, it can be used in triage to remove samples from well-detected families so that analysts can focus on the 38% unlabeled samples.

Originally FUD Families. Using their last report, AvClass outputs a family for 62.5% of the 600K originally FUD samples, which is in line with the overall labeling rate, indicating a similar fraction of well-known families among originally FUD samples. However, some families have larger fractions of originally FUD samples, and thus are harder to detect. Table 13 shows the top 10 families with at least 10K samples sorted by the ratio of originally FUD samples over all family samples. These include 6 families for Windows, 3 for Android, and one family of Microsoft Office macros. All of them have FUD ratios 6–16 times higher than the 0.59% average over all families with at least 10K samples.

5   Comparison with Telemetry

This section compares the VT file feed with the AV telemetry. We compare the total volume and percentage of malicious files, compute the intersection of malicious files, examine the family distribution, and measure which dataset observes malicious files faster.

Total and Malicious Volume. We first compare the total volume of both datasets over the one month when we were collecting all VT reports. Over that month, the VT file feed contains reports for 39.8M samples, while the telemetry contains events for 686.5M samples. Both numbers include all samples observed over that month in each dataset, regardless of the filetype, if the samples are old or new, and whether they are benign or malicious. Thus, the telemetry volume is 17 times larger than the VT file feed volume. The AV vendor has other file datasets available beyond the Windows telemetry (e.g., Android telemetry), thus its total file volume is even larger.

Over that month, the AV engine threw alerts for 1.9M malicious files in 905K devices, 0.3% of all samples seen in the telemetry over that month. In comparison, the VT file feed contains 14.8M samples with at least four detections (37.3%) and 17.5M with at least one detection (43.9%). Thus, the ratio of malicious files in the VT feed is 126–146 times larger than in the telemetry. This is likely due to two reasons. First, prior work has shown that AV telemetry is largely dominated by rare benign files, i.e., 94% of files in AV telemetry are observed only in one device and the ratio of benign to malicious such files is 80:1 [23]. Second, the VT file feed is likely biased towards malicious samples, as VT contributors may only submit suspicious samples to be analyzed, while avoiding to submit samples known to be benign.

Over the whole year, the AV engine detected 12.9M files as malicious. In comparison, the VT file feed contains 187.0M samples with at least four detections and 212.2M with at least one detection. Thus, over the course of the year the VT file feed observes 16–17 times more malicious files. Of the 12.9M detected files in the telemetry, 5.2M (40.3%) have extensions corresponding to peexe files (.exe, .dll, .sys, .cpx, .ocl, .scr), followed by .tmp temporary files (17.8%) and .lnk link files (9.7%).

Takeaway 8

While massive, the total VT file feed volume is 17 times lower than the Windows telemetry of a AV vendor. However, despite the much lower volume, the VT file feed contains 16–17 times more malware than the telemetry, making it a great source of malicious samples.

Intersection. We compute the intersection between both datasets over the whole year. Given the massive size of the telemetry (i.e., $> 10^8$ events), to make the query scale, we focus the intersection on malicious *peexe* files and ignore other filetypes and benign executables. Thus, we query the telemetry using the 151.7M *peexe* file hashes from the VT file feed with at least one detection. For each file hash found in the telemetry, we collect the anonymized identifiers of the devices where it was observed and the telemetry first seen time, i.e., the earliest time, within our collection period, a feed sample was queried by an endpoint to obtain its reputation.

The intersection contains 3.8M samples with at least one detection (1.8% of feed samples with one detection) and 2.2M (1.2%) with at least four detections. The small intersection indicates that the telemetry and the VT file feed observe largely disjoint sets of malicious samples. Prior work has observed that public and commercial threat intelligence feeds have small overlap [9,41]. However, those works focus on IP addresses [41] or work on APT-focused commercial TI feeds [9]. As far as we know, no prior work has checked the overlap between large (malicious) file hashes datasets. Our results show that even the largest (malicious) file hashes datasets are largely disjoint with minimal overlap. This is likely caused by a huge space of malicious samples of which each vendor only sees a small portion.

Of the 12.9M files detected as malicious by the AV vendor over the year, 11.9M (92.2%) are not observed in the VT file feed. These files are either never submitted to VT or their last VT report was before our collection start. Quantifying this requires querying the 11.9M files to VT which due to API restrictions is not possible. Instead, we estimate these figures by querying a subset of 1M randomly selected hashes. Only 10.9% of those are known to VT, while 89.1% have never been submitted. This shows that security vendors may only share a fraction of their malicious samples with VT. Sharing decisions by the AV vendor are transparent to us.

> **Takeaway 9**
> The telemetry and VT file feed observe largely disjoint sets of malicious samples (1.2%–1.8% of feed samples in common). Thus, even the largest file datasets only see a small portion of the whole space of malicious samples.

Family Distribution. Table 14 shows the top 10 families in the intersection sorted by number of telemetry devices where the samples of the family are observed. All these families are PUP. Instead, when we ranked families by number of samples observed in the VT file feed (*peexe* families in Table 5), the top families were dominated by virus and worm families. From the top 10 VT file feed families by number of samples, *vobfus* and *virlock* are the two families that affect most devices in the telemetry found on 25.9K and 3.3K devices, respectively, 1–2 orders of magnitude less devices than the families in Table 14. The remaining 8 families are ranked below the 1,000th position affecting each less than 2K devices. These results indicate that the top families in the VT file feed, i.e., those with the most samples submitted by contributors, may be biased towards highly-polymorphic families such as viruses and worms and may not correspond to the families that affect most user devices, which according to the telemetry are PUP families.

> **Takeaway 10**
> The top families by number of samples collected is biased towards highly polymorphic families such as viruses and worms, and may significantly differ from the top families by number of infected devices.

Observation Delay. The telemetry first seen timestamp for a sample, i.e., the earliest time within our collection period a feed sample was queried by an endpoint, is an upper bound on the earliest time the AV vendor observed the sample. For example, a sample first seen by the AV vendor in 2010 may appear in the telemetry subset we analyze as first queried on December 22nd, 2020. We calculate the delay to observe a sample as the VT first seen timestamp minus the telemetry first seen timestamp, but only for the 2.1M samples first observed by VT during our analysis period and that are in the intersection with the telemetry. Of those 2.1M samples, 2.5M (61%) are first observed by the telemetry (i.e., positive difference) while 816K (39%) are first observed by VT (i.e., negative difference). The median delay for VT to observe the sample is 4.4 h. Thus, real devices observe the sample a few hours earlier than VT. However, the mean delay is 21 days because 12% of these samples are first submitted to VT at least 3 months after they appear in the telemetry, compared to 3% being observed by VT 3 months earlier than in the telemetry. It is important to note that since the telemetry first seen is an upper bound for the AV vendor first seen, the VT delay may be actually larger.

> **Takeaway 11**
> Malicious samples are first seen a median of 4.4 hours earlier in the telemetry. Still, 39% of samples are first seen by VT before they are first seen in user devices. Thus, VT may provide useful early alerts to AV vendors.

6   Discussion

The section discusses the implications of our results for future works, limitations, threats to validity, and avenues for improvement.

Result Implications. Our results have implications for researchers analyzing the malware ecosystem. We show that the most popular Windows families widely differ between the VT file feed and the AV telemetry. Top families in the feed correspond to highly polymorphic malware such as viruses and worms. In contrast, families affecting most user devices are PUP. Thus, the most popular feed families may not be those that impact end users most, but rather those for which samples are easier to collect (e.g., due to their many polymorphic variants). Focusing only on the top feed families might ignore popular families that affect many user devices. Those families are also found in the VT file feed, but with lower volumes, so researchers may need to dive deeper into the feed beyond the top families.

Our results have implications for researchers that need to build malware datasets for ML approaches. The VT file feed is a great source for malware (and also benign) files, due to its large volume, filetype diversity, and freshness of samples. However, the diversity largely centers on Windows and to a smaller degree Android samples. For other platforms, even collecting samples for one year, would only provide a handful of families with at least 100 samples (e.g.,

52 for Mac OS and 37 for Linux), which we consider the minimum for training, validating, and testing ML family classification models.

Our results have implications for researchers building detection models on the VT file feed. Pendlebury et al. [28] argued that the goodware/malware ratio expected in ML testing datasets should be matched when training the model. They measured this ratio was 90/10 for AndroZoo [2]. Previous work has also shown that this ratio is roughly 99/1 in AV telemetry [23]. In contrast, we observe a ratio of nearly 50/50 for the VT file feed, indicating VT users are more likely to submit malicious samples. Accounting for this ratio is important for applying ML models on the VT file feed. To avoid temporal bias, Pendlebury et al. [28] also recommend that samples in the testing dataset have timestamps larger than any sample in the training dataset and that in every testing slot, all samples come from the same time window. For the VT file feed, this separation should use the VT first seen date because we show that 31% of the daily samples are re-submissions of older samples which may break these properties.

While the dynamics of detections labels have been studied before [8,44], our work is the first one that can analyze them on samples that are not selected apriori and re-scanned daily by the authors. This allows us to identify 600K originally FUD samples that initially escaped detection until multiple AV vendors realized their maliciousness a median of 7 days later (mean of 23.8 days). This raises the question of how many other malicious files may remain undetected in the feed.

Data Collection Issues. Longitudinal analyses often face unexpected data collection issues that create gaps in the temporal data sequence. Such issues prevented us from collecting VT reports on 39 days, most notably over 27 days between January 11th and February 7th, 2022. Thus, our dataset contains data for 326 days, rather than a whole year. We account for these gaps throughout the measurements, e.g., we do not provide yearly volume statistics, but provide daily statistics that exclude data gaps.

AV Telemetry Comparison. Our work shows that the VT file feed has little overlap with the telemetry of a large AV vendor and that the most popular families largely differ in both datasets. Results could differ for the telemetry of other AV vendors. However, we believe this is unlikely given the large size of both datasets. Furthermore, our results match those observed in smaller APT-focused file datasets [9] and in datasets of other malicious indicators such as IP addresses [41]. We believe the different results in this area indicate that feeds (even those that aggregate data from multiple other feeds) achieve limited coverage of indicators, thus highlighting the need for further aggregation and cooperation. It would be interesting to examine whether different AV vendors observe very different top malware families as well, but getting access to the telemetry of multiple AV vendors is challenging.

Family Labeling. Our malware labeling is based on AV labels processed by AvClass. Thus, it inherits the limitations of both the AV labels and the tool. For example, our results show that AV labels for document filetypes such as HTML and PDF often contain behaviors rather than family names. If the AV labels do not contain a family name, possibly because the AV vendors do not have a good definition of family for those filetypes, then AvCLASS cannot output a family. There are also cases where AvCLASS identifies as a family a token that is not a family (e.g., looks randomly generated). These may be due to new AV engines or changes to AV label format since AvCLASS was released. We will report them to the developers so that they can be addressed.

Filetype Identification. VT reports lack a unified filetype field. Instead, they provide the output of different filetype identification tools, which may not agree. To handle disagreements and minimize the number of samples without a filetype, we combine multiple filetype-related fields in the VT reports. Still, we cannot infer the filetype for 1.75% of samples indicating that further research on filetype identification is needed. Furthermore, filetype identification tools should output hierarchical filetypes allowing users to aggregate results as they prefer. For example, a DLL is also a PE executable and an APK is also a ZIP archive. Whether to count DLLs and APKs as their own filetypes or as part of their parent filetypes should be up to the user.

7   Related Work

Most related is the work by Ugarte-Pedrero et al. [40] that analyzes 172K PE executables that a large AV vendor collects through multiple sources on one day. In contrast, we examine one year of a file feed with 235M samples of multiple filetypes and compare it to the telemetry of a large AV vendor. Other works have performed large scale longitudinal malware analysis on Windows [7,22], Android [24,38], and Linux [3,12]. In contrast, our work examines malware for multiple platforms including Windows, Android, Linux, macOS, Microsoft Office macros, PDF documents, and Web content.

Detection labels such as those available in VT reports have been widely studied. Early works showed how different AV engines disagree on labels for the same sample [5,11]. Still, AV labels have been widely used to build training datasets and evaluate malware detection and clustering approaches (e.g., [5,6, 30]). Recent works have examined the dynamics of detection labels [8,44] and have proposed to replace the traditional threshold-based detection approach on the number of detections, which we use in this paper, with machine-learning models [34,39]. We plan to explore these approaches in future work.

8   Conclusions

We have characterized the VirusTotal file feed by analyzing 328M reports for 235M samples collected during one year, and have compared the feed with the telemetry of a large AV vendor. Among others, we show that despite having a volume 17 times lower than the AV telemetry, the VT file feed observes 8 times more malware. The feed is fresh with 69% of daily samples being new and samples appear a median of 4.4 h after they are seen in user devices. The feed is diverse containing 4.9K families with at least 100 samples. However, the diversity largely focuses on Windows and Android families. The AV telemetry and VT file feed observe largely disjoint sets of malicious samples (1.2%–1.8% overlap). We identify 600K originally FUD samples that have no detections on first scan, but are later considered malicious by at least 4 AV engines.

Acknowledgment. This work has been partially supported by the PRODIGY Project (TED2021-132464B-I00) funded by MCIN/AEI/10.13039/501100011033/ and EU NextGeneration funds.

References

1. Virustotal API 2.0 reference: File feed. http://developers.virustotal.com/v2.0/reference/file-feed
2. Allix, K., Bissyandé, T.F., Klein, J., Le Traon, Y.: AndroZoo: collecting millions of android apps for the research community. In: International Conference on Mining Software Repositories (2016)
3. Alrawi, O., et al.: The circle of life: a large-scale study of the IoT malware lifecycle. In: USENIX Security Symposium (2021)
4. Arp, D., Spreitzenbarth, M., Huebner, M., Gascon, H., Rieck, K.: Drebin: efficient and explainable detection of android malware in your pocket. In: Network and Distributed System Security (2014)
5. Bailey, M., Oberheide, J., Andersen, J., Mao, Z.M., Jahanian, F., Nazario, J.: Automated classification and analysis of internet malware. In: International Symposium on Recent Advances in Intrusion Detection (2007)
6. Bayer, U., Comparetti, P.M., Hlauschek, C., Kruegel, C., Kirda, E.: Scalable, behavior-based malware clustering. In: Network and Distributed System Security (2009)
7. Bayer, U., Habibi, I., Balzarotti, D., Kirda, E., Kruegel, C.: A view on current malware behaviors. In: LEET (2009)
8. Botacin, M., Ceschin, F., de Geus, P., Grégio, A.: We need to talk about antiviruses: challenges & pitfalls of av evaluations. Comput. Secur. **95**, 101859 (2020)
9. Bouwman, X., Griffioen, H., Egbers, J., Doerr, C., Klievink, B., Van Eeten, M.: A different cup of TI? The added value of commercial threat intelligence. In: USENIX Security Symposium (2020)
10. Buyukkayhan, A.S., Oprea, A., Li, Z., Robertson, W.K.: Lens on the endpoint: hunting for malicious software through endpoint data analysis. In: International Symposium on Research in Attacks, Intrusions, and Defenses (2017)
11. Canto, J., Dacier, M., Kirda, E., Leita, C.: Large scale malware collection: lessons learned. In: IEEE SRDS Workshop (2008)
12. Cozzi, E., Graziano, M., Fratantonio, Y., Balzarotti, D.: Understanding Linux malware. In: IEEE Symposium on Security and Privacy (2018)
13. Graziano, M., Canali, D., Bilge, L., Lanzi, A., Balzarotti, D.: Needles in a haystack: mining information from public dynamic analysis sandboxes for malware intelligence. In: USENIX Security Symposium (2015)
14. Huang, H., et al.: Android malware development on public malware scanning platforms: a large-scale data-driven study. In: International Conference on Big Data (2016)
15. Huang, W., Stokes, J.W.: MtNet: a multi-task neural network for dynamic malware classification. In: Detection of Intrusions and Malware, and Vulnerability Assessment (2016)

16. Hurier, M., et al.: Euphony: harmonious unification of cacophonous anti-virus vendor labels for android malware. In: IEEE/ACM International Conference on Mining Software Repositories (2017)
17. Jindal, C., Salls, C., Aghakhani, H., Long, K., Kruegel, C., Vigna, G.: Neurlux: dynamic malware analysis without feature engineering. In: Annual Computer Security Applications Conference (2019)
18. Kaczmarczyck, F., et al.: Spotlight: malware lead generation at scale. In: Annual Computer Security Applications Conference (2020)
19. Kotzias, P., Bilge, L., Caballero, J.: Measuring PUP prevalence and PUP distribution through pay-per-install services. In: USENIX Security Symposium (2016)
20. Kotzias, P., Caballero, J., Bilge, L.: How did that get in my phone? Unwanted app distribution on android devices. In: IEEE Symposium on Security and Privacy (2021)
21. Kotzias, P., Matic, S., Rivera, R., Caballero, J.: Certified PUP: abuse in authenticode code signing. In: ACM Conference on Computer and Communication Security (2015)
22. Lever, C., Kotzias, P., Balzarotti, D., Caballero, J., Antonakakis, M.: A lustrum of malware network communication: evolution and insights. In: IEEE Symposium on Security and Privacy (2017)
23. Li, B., Roundy, K., Gates, C., Vorobeychik, Y.: Large-scale identification of malicious singleton files. In: ACM Conference on Data and Application Security and Privacy (2017)
24. Lindorfer, M., Neugschwandtner, M., Weichselbaum, L., Fratantonio, Y., Van Der Veen, V., Platzer, C.: Andrubis-1,000,000 apps later: a view on current android malware behaviors. In: International Workshop on Building Analysis Datasets and Gathering Experience Returns for Security (2014)
25. Maffia, L., Nisi, D., Kotzias, P., Lagorio, G., Aonzo, S., Balzarotti, D.: Longitudinal study of the prevalence of malware evasive techniques. arXiv preprint arXiv:2112.11289 (2021)
26. Mantovani, A., Aonzo, S., Ugarte-Pedrero, X., Merlo, A., Balzarotti, D.: Prevalence and impact of low-entropy packing schemes in the malware ecosystem. In: Network and Distributed Systems Security Symposium (2020)
27. Masri, R., Aldwairi, M.: Automated malicious advertisement detection using VirusTotal, UrlVoid, and TrendMicro. In: International Conference on Information and Communication Systems (2017)
28. Pendlebury, F., Pierazzi, F., Jordaney, R., Kinder, J., Cavallaro, L.: Tesseract: eliminating experimental bias in malware classification across space and time. In: USENIX Security Symposium (2019)
29. Peng, P., Yang, L., Song, L., Wang, G.: Opening the blackbox of VirusTotal: analyzing online phishing scan engines. In: Internet Measurement Conference (2019)
30. Perdisci, R., Lee, W., Feamster, N.: Behavioral clustering of HTTP-based malware and signature generation using malicious network traces. In: USENIX Symposium on Networked Systems Design and Implementation (2010)
31. Pontello, M.: TrID - File Identifier (2021). http://mark0.net/soft-trid-e.html
32. Raff, E., Barker, J., Sylvester, J., Brandon, R., Catanzaro, B., Nicholas, C.K.: Malware detection by eating a whole EXE. In: Workshops at the AAAI Conference on Artificial Intelligence (2018)
33. Rieck, K., Holz, T., Willems, C., Düssel, P., Laskov, P.: Learning and classification of malware behavior. In: Detection of Intrusions and Malware, and Vulnerability Assessment (2008)

34. Salem, A., Banescu, S., Pretschner, A.: Maat: automatically analyzing VirusTotal for accurate labeling and effective malware detection. ACM Trans. Privacy Secur. **24**(4), 1–35 (2021)
35. Sebastian, M., Rivera, R., Kotzias, P., Caballero, J.: AVClass: a tool for massive malware labeling. In: Research in Attacks, Intrusions, and Defenses (2016)
36. Sebastián, S., Caballero, J.: AVClass2: massive malware tag extraction from AV labels. In: Annual Computer Security Applications Conference (2020)
37. Smutz, C., Stavrou, A.: Malicious PDF detection using metadata and structural features. In: Annual Computer Security Applications Conference (2012)
38. Suarez-Tangil, G., Stringhini, G.: Eight years of rider measurement in the android malware ecosystem. IEEE Trans. Depend. Secure Comput. (2020)
39. Thirumuruganathan, S., Nabeel, M., Choo, E., Khalil, I., Yu, T.: SIRAJ: a unified framework for aggregation of malicious entity detectors. In: IEEE Symposium on Security and Privacy (2022)
40. Ugarte-Pedrero, X., Graziano, M., Balzarotti, D.: A close look at a daily dataset of malware samples. ACM Trans. Privacy Secur. **22**(1), 1–30 (2019)
41. Li, V.G., Dunn, M., Pearce, P., McCoy, D., Voelker, G.M., Savage, S.: Reading the Tea leaves: a comparative analysis of threat intelligence. In: USENIX Security Symposium (2019)
42. VirusTotal. http://www.virustotal.com/
43. Yuan, L.-P., Wenjun, H., Ting, Yu., Liu, P., Zhu, S.: Towards large-scale hunting for android negative-day malware. In: International Symposium on Research in Attacks, Intrusions and Defenses (2019)
44. Zhu, S., et al.: Measuring and modeling the label dynamics of online anti-malware engines. In: USENIX Security Symposium (2020)

Attackers as Instructors: Using Container Isolation to Reduce Risk and Understand Vulnerabilities

Yunsen Lei[1(✉)], Julian P. Lanson[1], Craig A. Shue[1], and Timothy W. Wood[2]

[1] Worcester Polytechnic Institute, Worcester, MA, USA
{ylei3,jplanson,cshue}@wpi.edu
[2] George Washington University, Washington, D.C., USA
timwood@gwu.edu

Abstract. To achieve economies of scale, popular Internet destinations concurrently serve hundreds or thousands of users on shared physical infrastructure. This resource sharing enables attacks that misuse permissions and affect other users. Our work uses containerization to create "single-use servers" which are dynamically instantiated and tailored for each user's permissions. This isolates users and eliminates attacker persistence. Further, it simplifies analysis, allowing the fusion of logs to help defenders localize vulnerabilities associated with security incidents. We thus mitigate attacks and convert them into debugging traces to aid remediation. We evaluate the approach using three systems, including the popular WordPress content management system. It eliminates attacker persistence, propagation, and permission misuse. It has low CPU and latency costs and requires linear memory consumption, which we reduce with a customized page merging technique.

1 Introduction

Internet servers are designed to handle many clients simultaneously. These servers use multiple processes or threads of execution to balance requests and make effective use of computing resources. Unfortunately, this model intermingles processing from many clients within a single execution context. When these servers have security defects, attackers can exploit the vulnerabilities to gain unauthorized access, modify the server's content, and harm other current or future users [31].

Exacerbating this problem, when a server accesses other resources, such as databases, it is often configured with a super-set of all privileges associated with the server's users. This can lead to "confused deputy" attacks [13], wherein an adversary exploits a vulnerability to cause an application server to misuse its authority when interacting with a resource provider. SQL injection attacks, which are estimated to be used in nearly two-thirds of all web attacks [3], are a common form of confused deputy attack.

In this work, we propose a "single-user server" model where each incoming client gets directed to its own isolated container. We explore a set of research

D. Gruss et al. (Eds.): DIMVA 2023, LNCS 13959, pp. 177–197, 2023.
https://doi.org/10.1007/978-3-031-35504-2_9

questions: *How can this single-use server model limit attack propagation, persistence, and privilege escalation? Can containerization provide low enough overheads to support a large number of concurrent users? To what extent can we improve the resource consumption of the approach? What impact can the single-use server model have on attack reconstruction and analysis?*

The first two research questions lead to novel contributions in container management and access control. Our single-use server model places every application server in its own Docker container with permissions tailored to the associated end user. When a client first connects to a server, it has anonymous user privileges and tightly constrained access to backend resources, such as a database. When a user authenticates, our approach automatically alters the permissions associated with the container to match the privileges associated with the authenticated user. Since the permissions for each application server and container are tailored, they do not have the elevated privileges necessary to enact a confused deputy attack. The container approach provides isolation and the destruction of a container upon the client's disconnection eliminates attacker persistence.

The third research question leads to novel contributions to memory deduplication. Our approach, called Focused Kernel Same-page Merging (FKSM), actively merges two container's processes if they run the same programs.

The final research question leads to novel contributions in attack analysis and localization. To detect access violations, we create monitoring infrastructure for communication between clients and the application server as well as between the application server and any back-end resources. This monitoring also enables forensic reconstruction of attacks. In this work, we:

- **Design a Single-Use Server (SuS) approach** that includes the components needed for authentication, container management, and the collection of forensics for arbitrary applications (Sect. 3). *Our design improves security by cleanly separating the untrusted execution environments for individual users from each other and from the control plane that routes, authorizes, and monitors requests.*
- **Implement a Single-Use Web Server** using a novel combination of lightweight containers and network middleboxes to support three web application services, including the popular WordPress platform (Sect. 4). *Our implementation demonstrates that applications can be ported with minimal codebase modifications. We also enable fine-grained permissions to be safely enforced by proxy middleboxes. We further develop a memory deduplication approach that saves 26% of memory for each container while the merge time is only a fraction of the state-of-the-art UKSM [40] approach.*
- **Evaluate the security and performance** of our SuS implementation (Sects. 5 and 6). *We find the containerization approach prevents several exploits against vulnerable versions of WordPress without requiring application software patches. It incurs less than 5% CPU overhead, needs only 2GB of RAM to run 100 concurrent containers, and shows only a 20% increase in response time when running 100 concurrent containers.*
- **Reconstruct attacker steps** by leveraging the per-user logging enabled by SuS (Sect. 5.2). *When exploring a known CVE attack on our SuS WordPress*

system, we find that a back-tracing workflow can quickly localize the search space for debugging and remediation, reducing the search space from thousands of files to only two functions.

2 Background and Related Work

In this section, we review work in the most related areas and discuss how our approach is different from various perspectives.

Security through Isolation: Parno et al. proposed CLAMP to protect LAMP-stack websites [27]. CLAMP assigns individual users to isolated VMs running copies of the web server code. Users can upgrade their VM's permissions via a separate, trusted authentication portal. Unfortunately, CLAMP provides only 42% of the performance of native operations. The CLAMP authors acknowledged significant impediments to the practical deployment of such a system and did not complete an analysis of VM start-up on normal operation, citing the significant overhead of VM start-up and limitations of delta virtualization. In contrast, in our work, we designed and implemented customized memory improvements and performed an end-to-end evaluation, including on-demand server generation. Our result shows that SuS incurs modest overheads and achieves greater scalability. Taylor [36] introduces a software-defined networking (SDN) controller to demultiplex users' traffic and guide their packets to isolated VMs. The Taylor work lacks the resource restriction component present in CLAMP, but adds attack attribution. Our SuS model improves performance and scalability; further, it uses log fusion to reconstruct attack steps, which was not previously explored.

Radiatus, by Cheng et al. [6], builds off CLAMP and introduces more stringent security measures. The result of such a design is a web development framework that requires developers to use their API to create an application. Porting an existing application to use Radiatus thus requires re-implementation. Our SuS model aims to provide strong security isolation and can be deployed on widely used web applications like WordPress and HotCRP with minimal code base modification (≤ 50 lines of code).

Lighter Weight Virtualization: Some Internet services use lightweight virtualization like containers to facilitate fast deployment, fine-grained scaling, and component failure isolation [34,35]. Prior work has also sought ways to reduce the resource consumption and cost of starting and running applications with these technologies [2,15]. Serverless computing platforms benefit from these lightweight virtualization technologies. SuS is different from the serverless model. Our work takes a user-based view of the application and constrains the user's behavior based on the functions and resources the user is supposed to access. We focus on security and forensics aspects.

Memory Deduplication: Kernel Same-page Merging (KSM) on Linux allows applications to share identical pages by comparing the page content. Previous work improves KSM in terms of scanning speed and resource utilization [11, 33,41]. UKSM [40] improves KSM by prioritizing statically-duplicated memory regions and reducing computational cost through Adaptive Partial Hashing [12].

Our merging approach differs from these existing works in the way duplicate pages are identified. We compare our FKSM with UKSM on deduplication speed and effectiveness to show the benefit of active and strategic scanning.

Forensic Analysis of Exploits: Data records can help defenders understand, analyze and replay past events that are related to attacks. Dunlap et al. [8] proposed Revirt, which uses checkpoint logging and roll-forward recovery to replay entire attack events. To facilitate the human understanding of collected forensics, researchers have proposed different approaches [7,10,32] to visualize the data. In our work, we focus on constructing execution traces in an informed way that leverages per-user isolation, facilitating visualization integration.

3 An Untrusted Application Server Design

Application servers are complex, making them ripe for attack. We create monitoring and protection components so that attacks become valuable learning opportunities for defenders to improve software without negative outcomes.

3.1 Threat Model

The SuS model is a server-side defense system aimed at preventing adversaries 1) from successfully executing any backend request above their intended privilege level and 2) from making changes to server files that would enable them to attack other users. We assume that adversaries can only access the server program's host machine via network communication and that they will attempt exploits via the packet payloads in the server program's communication protocol.

We assume the application server within a single-use container can be exploited and adversary-controlled. The adversary may arbitrarily control one or more clients and containers. We assume that the container facilitates access to information stored in one or more backend resources (e.g., in databases or file shares) but that it otherwise only stores per-user session state. For the defender, we assume that they leverage the SuS capability to configure the levels of access based on their applications and different user roles. Such configuration exercises a least-privilege principle, helping defenders mitigate exploitation against unknown vulnerabilities.

We exclude attacks that cause privileged users to misuse their legitimate privileges, such as social engineering or cross-site request forgery. Similarly, we exclude attacks against our trusted computing base (TCB), which includes the operating system, the back-end resource servers, and the SuS infrastructure components themselves (such as middleboxes and container managers). While we evaluate container scalability and performance, we exclude flooding-based denial-of-service attacks and assume defenders employ current best practices. While a trusted kernel is a common threat model assumption, and one we use as well, we recognize that efforts to escape a container and elevate privileges are possible. Given the importance of kernels and containers, we anticipate other continued efforts to improve and protect them. The scope of the SuS approach is to explore the feasibility of using lightweight virtualization to run server instances that are tailored to a single user. Our approach could be used with other lightweight isolation and virtualization technologies as needed.

3.2   Design Components

Our SuS model assumes that each server instance will support only a single client, although this could be extended to enable a container to serve a group of related users, albeit with no protection between them. We will place each application server in a separate container and examine each container for indications of compromise (IOCs) that merit further analysis. Any container that lacks an IOC is deleted once it is no longer needed by a client. Figure 1 shows the following SuS model components. We now describe these components in detail.

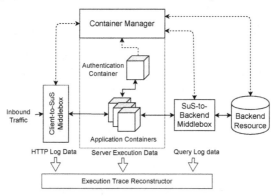

Fig. 1. Design overview of Single-use Server architecture. The middleboxes, authentication, and management components coordinate to provision SuS containers and assign them to clients, provide a means of upgrading privileges to the backend resource, and contain and analyze exploited application servers.

3.3   Application Containers

We place each application server in its own isolated execution container. These containers are instantiated by cloning an existing server. Before the container interacts with a client, we consider the container to be "pristine." While it is pristine, the container can be trusted since it is unreachable by an adversary. Once a container interacts with a client, we consider the container "contaminated" and inherently untrustworthy.

3.4   Container Management

The centralized Container Manager mostly performs management tasks including server instance configuration, container provisioning, reclamation, or freezing (if an attack is detected). Because the Container Manager knows which pair of processes are created from the same resource, it also communicates with the FKSM kernel module to initiate page deduplication.

3.5   Authentication Container and Permissions

Many application servers must identify the user associated with a given client. In our model, we cannot rely on the untrusted application to accurately report the client's authentication status. Instead, all user authentication is handled by a trusted Authentication Container, which has a minimal code base that can be more easily verified and protected. This separation of roles is somewhat similar to the Kerberos authentication model [17].

The Container Manager communicates with the Authentication Container to adjust container permissions. If the Authentication Container confirms a client's identity, the Container Manager increases privileges in the SuS-to-backend middlebox and backend infrastructure to reflect the new user permissions. In essence, the SuS container gains only the privileges associated with the connected user.

Our privilege model differs from traditional web server configurations. An application container does not need a superset of users privileges during the configuration (e.g., WordPress installation recommends granting all privileges in the WordPress database). Such a configuration has the potential risk of letting adversaries ultimately control the entire WordPress database in case of an exploited WordPress instance. In our SuS model, the same exploitation is limited to the database privileges of the account associated with the exploited container. In other words, the adversary may issue queries, but the queries will only succeed if they can be performed within the limited permissions associated with the container. We treat database server privilege errors as evidence of compromise.

3.6   Client-Side Demultiplexing and Forwarding

Our Client-to-SuS middlebox acts as a load balancer that demultiplexes clients and directs each to a separate SuS container and as a proxy that handles all encryption functionality. This keeps cryptographic keys out of the untrusted SuS container environment while letting the middlebox vet unencrypted data. It controls access and blocks client communication in the event of an access violation. It logs network traffic for forensic reconstruction. This allows defenders to pinpoint the client messages that preceded the violation, potentially revealing the vulnerability exploited in the SuS container.

3.7   Guarding Backend Resources

The narrowly-tailored permissions the Container Manager configures for backend resources solve many Confused Deputy attacks. However, a SuS-to-backend middlebox provides fine-grained restrictions that some backend implementations cannot support. For example, an authenticated user should have UPDATE privileges to the application's users table so the user can change the associated email address or password. However, that privilege should be limited to certain rows, such as the rows for which the column USERID matches the authenticated user's identifier. The SuS-to-backend middlebox must be protocol-aware to perform resource access control effectively. We designed the system to easily swap between backend modules for protocols such as SQL and NFS. The modules in the middlebox must implement specific API functions that (1) parse and conditionally modify resource requests and (2) detect permission violations in request responses. The middlebox also observes any errors or responses from the backend resource and informs the Container Manager to act accordingly. The middlebox logs the communication for incident analysis of any attacks and it prevents access between potentially compromised SuS containers and the backend resources.

3.8   Constructing Execution Events

The client-side and backend-side traffic are logged by the corresponding middleboxes and associated with users' identities. With our design, this traffic is automatically separated to represent a single user's interaction with the service. While helpful, the network traffic alone is insufficient because it lacks insight into the user's interaction with the application. In Sect. 4.6, we describe the server profiling component that provides such detail. Since SuS logs the per-user server instances, it has a significant advantage in fusing and reconstructing logged data to facilitate the understanding of the provenance and the impact of an incident. We discuss the implementation of log integration in Sect. 4.7.

4   Implementation

To provide concrete examples and show evidence of generalization of our SuS design, we create implementations using three different Web applications: 1) WordPress [39], a popular Internet content management system estimated to be used in over 35% of all websites on the Internet [37]; 2) HotCRP [14], a system for managing paper submission and peer reviews for conferences; and 3) an anonymized learning management system (LMS) used in our organization for the administration and delivery of class materials. For simplicity, we focus on WordPress and simply describe where HotCRP and our LMS applications differ.

All three applications require a web server, a PHP runtime, and a database server. Since PHP is used in over 78% of popular websites [38], we focus on PHP web applications. We use the popular Docker container system to implement our containers. We test multiple versions of the WordPress software to measure the impact of the single-use server approach on attacks against versions of WordPress with known vulnerabilities. In the remainder of this section, we describe the implementation details of our SuS containers, the Container Manager, the Authentication Container, the middleboxes, and our protocols.

4.1   Container Configuration

We build a Docker container image through a "Dockerfile" for each application we need to protect in a SuS container. The application container is configured under a private network which is created through Docker's command line interface. We assign each container a unique IP and expose the necessary ports.

Under their standard deployment, most web applications must be configured to communicate external components such as server applications and resource databases. The SuS model separates server instances but does not fundamentally interrupt the data flow. Therefore, a similar configuration effort is required. Using WordPress, HotCRP, and our LMS as case studies, we find that only minor modifications are required and mainly involve the following aspects:

Authentication: The authentication logic must relocate from the untrusted container into the Authentication Container. For all three applications, we rewrite the login URL and direct the user to their assigned SuS container after successful authentication. For HotCRP, users must log in for most features and be directed to a pristine container. Our LMS application was configured as a relying party associated with a single-sign-on identity provider. Therefore, the redirection URL is encoded with the parsed SSO response.

Shared Resources: The SuS-to-Backend Middlebox (Sect. 4.5) can support backend resources such as databases. Some services, like HotCRP, use a mail server for message transmission. This can be handled via an external server or a shared service (which itself could be in the SuS system). For simplicity, we simply use a mail server on the container host in our experiments.

4.2   Container Manager

The Container Manager creates a thread to maintain a pool of available containers for new clients. Our pooling strategy hides latency by ensuring a container is ready when a client arrives. When stopped, the Container Manager terminates all threads and containers and removes container credentials from the database. We use a startup script as the entry point that setup control arguments for the rest of the container processes. After parameter configuration, the script then fires up the web application. The scripts receive these arguments from the Container Manager as part of the container startup process. An internal control manager handles the generation of each container's control arguments (including IP address, database credential with minimal privilege, PHP settings, etc.).

Optimizing memory usage for SuS is important. Our memory deduplicator is implemented as a kernel module in Linux and a userspace component which is part of the Container Manager. The communication between the kernel and userspace is achieved through a `ioctl` call in which the manager passes a pair of in-container process IDs that requires merging. Container processes' IDs are obtained by intercepting the Docker event interface and examining the corresponding `cgroup` directory on a container startup event. After receiving the process IDs, our kernel module's callback function scans the processes' pages. Before merging, we compute and store page metadata in a two-layered hashmap. This structure combines xxHash checksums with Blake2b checksums for each page to perform faster merge comparisons. The first layer is indexed off xxHash's first bytes, and the second layer stores the full xxHash and points to a red-black tree indexed off the Blake2b hash. The mergeable pages between two processes are first maintained in a link list and then merged using existing kernel functions.

4.3   Authentication Container

The Authentication Container operates a web page that prompts clients for credentials. Upon receiving the user's credentials, it tries to validate them and notifies the Container Manager whether the client has a new role. The Container Manager then accesses the database and appropriately upgrades the privileges of the account associated with the client's assigned SuS container. The Authentication Container finally redirects the user back to a specific URL on the appropriate SuS container. That URL encodes data that allows the script to set authentication cookies to set the user identity in the application.

This authentication model requires that the two redirect messages be cryptographically validated. We encode nonces and a message authentication code in the redirect messages to ensure the authenticity of all passed parameters. We derive the keys using information preconfigured by the Container Manager. This approach allows the Authentication Container to statelessly validate messages from any SuS container without requiring an interaction with the Container Manager. We omit the details of these messages for brevity.

4.4   Client-to-SuS Middlebox

The main task for the client-to-SuS container middlebox is determining the appropriate container for each client. We embed the user information in an HTTP cookie called `SUS_DEMULTIPLEX_COOKIE` to perform the client demultiplexing. We implement the middlebox using Python's `asnycio` library and use its API functions to handle the TLS termination. The middlebox can thus parse the HTTP request to extract the user information within the `Cookie` header. After parsing the HTTP header, it updates an internal mapping structure with the relation between the user information and its assigned SuS container IP. The validity of the `SUS_DEMULTIPLEX_COOKIE` determines whether a new container request is needed. After a SuS container is purged for inactivity, the middlebox removes the corresponding internal mapping. Upon receiving the first response from the SuS container, the middlebox rewrites the HTTP response in a `Set-Cookie` field with appropriate cookie value and expiration to ensure the client sends subsequent requests with the cookie value. It then encrypts the response message and sends it back to the client.

We also implement a second cookie named `SUS_LOG_COOKIE`, which is used only on the server side to uniquely name each request for logging and event reconstruction purposes (See Sect. 4.6 and Sect. 4.7).

4.5   SuS-to-Backend Middlebox

The SuS architecture ensures that the database can enforce the table-level constraint by itself. Fine-grained query scoping requires configuration similar to prior work [16]. In our work, we create a proxy middlebox between the SuS containers and the MySQL database. As described in Sect. 3.7, one task that middlebox performs is query scoping, which limits table access to certain rows.

We define a *ResourceRestrictTable* that maps the tuple *(Role, Resource, Access Type)* to an *Access Predicate*. An Access Predicate is an extra limitation that can be applied to the query by appending it to the query's `WHERE` clause or an assertion to check the presence of a specific row selector in the `WHERE` clause. The middlebox also maintains a *UserContext* dictionary that maps container IP to user information (e.g., user_id, role). For each database query, the middlebox first retrieves the user's role based on the container IP. Then it extracts the resource (table) and access type (e.g., `SELECT`, `INSERT`). The middlebox retrieves the Access Predicate using the above three values. An access predicate may have a variable, as in the case `ID = :user_id`. In this case, the middlebox inserts the corresponding value from the UserContext dictionary entry before appending it to the query. The modified query is then sent to the MySQL database, and the response is forwarded to the container from which the query originated as usual. This silently restricts the data available to each user. The middlebox monitors and logs MySQL server responses for permission violations and regards any such error messages as an indication that the container has been compromised. Upon detecting such an error, the middlebox issues a request to freeze the container.

4.6   Tracing Application Execution

We implement a PHP extension that leverages request hooks to mark the start and end of a request and log important contexts such as URLs and cookies. The `SUS_LOG_COOKIE` cookie (inserted by the Client-to-SuS middlebox as mentioned in Sect. 4.4) is extracted at the `request_start` hook function. In addition to the request information, we also leverage the function execution hook to record function execution information, including the function name, the function call site (the file and line at which the function is called), and the function's parameter values. In this hook, a `SUS_LOG_COOKIE` value will be propagated if the function execution is part of the request handling. Because PHP handles requests synchronously, this propagation is scoped by the `request_start` and `request_end` hook functions.

The profiling extension is application-independent and can be loaded and unloaded through the PHP runtime's configuration file when PHP processes start. We found that accessing a complete list of function parameter values can incur significant overhead. Therefore, we only obtain the first three elements' values for composite-type parameters. In addition, we limit the parameter tracing depth when a composite-type parameter contains other composite-types.

Since the profiling extension runs within each SuS container, the profiling data may be tainted. An attacker with control of the SuS container may manipulate the profiling extension to provide false data. We leverage Linux's `auditd` from outside the container to implement rules that monitor `ptrace` and accesses to PHP's configuration directory. These rules can effectively detect an attacker's attempt to subvert the profiling modules through code injection and module replacement. Previous work explores syscall semantic reconstruction for interpreted program [5,18]. Tracing syscalls from outside the container can enable legitimacy estimates of the PHP execution trace. Mismatches between the syscall traces and PHP traces could themselves be indicators of container compromise.

4.7   Integrating Execution Traces

As mentioned in Sect. 3.8, logs from different users can easily be separated because of the single-use design. Our system generates: (1) an HTTP log from the Client-to-SuS middlebox, (2) PHP execution logs from profiling modules, and (3) a resource query log from the SuS-to-backend middlebox. These logs depict an interaction from different perspectives and, when integrated, constitute an execution trace of the whole event.

The first step in constructing the trace is parsing unstructured PHP logs into per-request call graphs. We implement a syntactical parser based on `php-ast` [30] to locate the user-defined function's definition (a script that defines the function and the line ranges of the implementation) and functions that will be called in the global scope. This statically-learned information is combined with our profiling data to construct the graph. The former allows us to determine the calling relation between two function log entries (e.g., whether function A is called within the block of function B). The latter allows us to identify the root node of a chain of function execution. The resulting request call graph is essentially a tree that starts with a root node named `RINIT` describing the request. Each child of the `RINIT` node is a global scope function followed by subsequent function calls and ends with PHP sink functions such as `mysqli_query`. To link the HTTP log to the request call graph, we only need to match the `SUS_LOG_COOKIE` cookie.

Table 1. CVEs and defenses considered in our security evaluation.

Category	CVE	Vulnerability Description	SuS Attack Mitigation
Single-user Instances	2012-3578 [20]	Input type validation failure enables script file upload and execution of arbitrary SQL queries and database credential leak	At SuS container startup, each `wp-config` PHP file is written with a database user of minimum privilege
Table-Level Privilege Constraints	2021-24182 [24]	Union-based SQL injection on `wp_tutor_quiz_question_answers`	Deny access to sensitive tables for unauthenticated or limited users (e.g., student roles)
	2021-24183 [25]	Union-based SQL injection on `wp_tutor_quiz_question`	
	2018-19207 [1]	Allows update to `wp_setting` table to register new admin account	Limit update access to the `wp_setting` table to administrator users
Row-Level Query Scoping	2019-9879 [21]	Privilege escalation exploit allows unauthenticated user to register new admin user	Query scoping prevents `wp_capability` used as the row selector when updating `wp_usermeta`
	2020-13693 [23]	Privilege-escalation exploit allows unauthenticated user to change `wp_usermeta` table to register with `bbp_keymaster` role	
	2019-9880 [22]	Allows unauthenticated user to retrieve all user information in `wp_user` table	Query scoping adds row selector to limit user access to only their own data
	2009-2762 [19]	Input validation failure allows reset of administrator's password for account hijack or account-level denial-of-service	

This approach allows us to link accurately even with the server application's URL rewriting. To link the call graph with the resource log query, we match the query string with the PHP function's parameter.

5   Security Evaluation

To evaluate the security benefits of the single-use server, we first consider the attacker's goal and common techniques in exploiting the confused deputy. Normally, attackers aim to access resources beyond what the application is designed for or what a user is allowed. This often requires the attacker to inject specific queries or misuse existing queries in the application code. For the injection case (the first three cases in Table 1), we consider two common attack vectors: file uploads and SQL injection. For the query misuse cases, we examine a set of attacks of this type (The fourth to eighth cases in Table 1).

We explore SuS effectiveness against attacks by classifying these exploits based on the SuS feature that stops or mitigates the attack. For each vulnerability, we apply an exploit to a test environment, both with our SuS model and without it. With the SuS model, the defender is required to configure the enforcement policies, while the control uses a shared server of the same software.

5.1   Evaluation: Real-Word Vulnerabilities

Single-user Instances For a server deployed using SuS model, file upload vulnerabilities are naturally mitigated because any uploaded script is only accessible within the attacker's container. Further, the uploaded script can only execute database queries within the limited permissions granted to that container (e.g., credentials saved in files like wp-config.php). We used CVE-2012-3578 [20] to evaluate SuS's effectiveness. As expected, the attack was successful in the shared server scenario. Our script, which aimed to delete WordPress accounts, failed in the SuS model since the container lacked the necessary permissions.

Table-Level Privilege Constraints Since each SuS container is configured with a unique database user account, we can configure different table access privileges based on the identity associated with the client. This prevents confused-deputy attacks since it eliminates privileged access that must be granted on a shared sever. For SQL injection attacks, queries which are manipulated to access any sensitive tables, such as mysql.user, are denied. In Table 1, we select three representative CVEs and show how SuS is configured to address these attacks.

Row-Level Query Scoping This class includes attacks in which a malicious user takes advantage of permissions that they were intentionally given in order to access or modify data that is disallowed by the security policy. This typically happens when the backend resource's native access control system is too coarse-grained to properly implement the desired policy. Our SuS-to-Backend middlebox enforces access control at the row level. We evaluate such a control's effectiveness through 4 different CVEs, as shown in Table 1.

5.2   Case Study: Exploring Execution Traces

While SuS can block the requests that trigger security exceptions, this alone does not help a security analyst to identify the root cause of an exploit. To illustrate how SuS logging can guide the process of localizing vulnerabilities, we consider a case study. We explore the *GdprOptions* vulnerability (CVE-2018-19207) and analyze the data. Using the trace construction workflow from Sect. 4.7, we construct a graph of relevant events from each HTTP request and SQL interaction. As mentioned above, the *GdprOptions* vulnerability is prevented by our SuS model because the UPDATE query needed to change the WordPress site settings exceeds the permissions associated with the client. This allows us to prune our analysis to graphs that contain the denied update query. The result, depicted in Fig. 2, shows progression from an HTTP POST request (the first block) to /wp-admin/admin-ajax.php, and a series of PHP function calls that end with a denied SQL query (the last block).

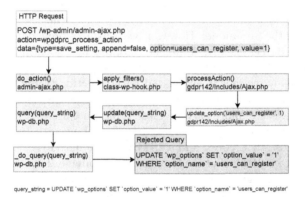

Fig. 2. The pruned trace shows how the HTTP request (yellow) is handled by PHP functions (blue) and leads to the rejected SQL query (red). For readability, we use the string "query_string" as shorthand to represent the full SQL query that appears as parameters in the PHP nodes. (Color figure online)

In the trace, the PHP execution starts in the WordPress codebase and eventually enters the code base of a plugin (gdpr142/Includes/Ajax.php). After execution of the processAction, the query is prepared and processed by a sequence of query-related functions until it is issued by the _do_query function. Since the query is simply passed through those helper functions unmodified, it suggests that the issue originates in or near the processAction or update_option functions. The documented patch to the vulnerability confirmed that the processAction was indeed the cause [26]. With such data, a defender can remedy the issue by patching the software or removing the plugin.

In a non-SuS system, this exploit might not be noticed for days or weeks, at which point logs may be overwhelming to analyze, and there will be no clear trail back to the request that triggered the exploit. In contrast, SuS's user separation allows the query to be rejected, signaling the need for immediate analysis. In this particular example, there are a total of 802 functions and 12 script files that are accessed between the HTTP request and the SQL query (assuming the analyst can identify the malicious request). SuS allows an analyst to quickly narrow the potential cause of the vulnerability to only two functions (`processAction` and `update_option`) within a single file (`gdpr142/Includes/Ajax.php`), out of the 1000+ PHP scripts in a WordPress installation.

We also conducted the same attack reconstruction analysis for other CVEs described in Table 1 as well. The log reduction benefit applies to other CVEs too; reconstructed traces average 18 functions with an average total of 1188 lines.

6   Performance Evaluation

Scalability and performance are key considerations for the SuS model. Since containers can use shared read-only mount points, and the remaining temporary file write space is needed in shared servers anyway, disk utilization is not a significant concern for the SuS model. However, we must explore what additional CPU and memory resources, if any, would be required by allocating separate server instances for each client and isolating them in separate containers. We must also explore the latency overheads associated with directing traffic to the appropriate container, logging its interactions, and enforcing permissions associated with those containers. We explore each of these topics in turn.

All our SuS containers run within a virtual machine with 16 GB of RAM, an allocation of 4 host CPU cores, and a 40 GB virtual hard drive. The VM runs on a physical host with 192 GB of RAM, 20 cores running at 2.20GHz, and 21 TB of hard drive space, configured with RAID. Our containers are not configured to use or enforce any CPU or memory limits. This configuration allows a comparison with the performance results associated with CLAMP [27].

6.1   RAM Usage

Before comparing memory usage between the SuS and the shared server model, we explore multiple server configuration options to ensure a fair comparison. We use WordPress as an example to show the impact of these options and determine the best choice for each model. We first configured different web servers using PHP with a static pool. For the shared and SuS server, the PHP worker pool size is set to the maximum number of concurrent users and one per container, respectively. The memory usage is calculated through Linux's `free` command. In the shaded portion of Table 2, we show the memory usage ratio (i.e., $\frac{SuS}{Shared}$) of the SuS model versus the shared server model. Our experiment shows that the shared server only uses a subset of the configured workers and achieves memory sharing that the SuS model does not.

For subsequent experiments, we select `nginx` with PHP-FPM as the default configuration for SuS because its relative lightweight and server popularity. Using

Table 2. The ratio of memory ($\frac{SuS}{Shared}$) used by WordPress in the SuS model verses the shared server model across varying numbers of concurrent users in 10 trials of experiments. The shaded results omit copy-on-write sharing while FKSM uses such page sharing in the kernel.

Server Configuration	Concurrent Users	10	25	50	100
Apache PHP-FPM	original approach	3.94	5.01	6.97	8.1
	Our FKSM	2.72	3.37	4.71	5.39
Nginx PHP-FPM	original approach	4.13	5.08	5.67	7.37
	Our FKSM	2.85	3.47	3.86	5.0
Lighttpd PHP-FPM	original approach	4.02	4.96	5.48	6.9
	Our FKSM	2.89	3.33	4.65	5.59

PHP-FPM with a single PHP worker, we pre-spawn a fixed number of SuS containers and use web clients to interact with the servers. We measure active containers' memory usage while serving pages to clients. In Table 3, we see that the per-container memory usage decreases as the number of concurrent clients increases. This is likely the result of amortizing fixed costs.

Table 3. Per-container memory usage (in MiB) with active clients across three applications using `nginx` and PHP-FPM. Results averaged over 10 trials.

Concurrent Users					
SuS Application		10	25	50	100
memory usage in MiB	WordPress	31.25	30.02	29.21	28.37
	HotCRP	27.80	27.71	27.55	27.16
	LMS	23.12	22.90	21. 14	20.70

The application server's basic properties play a significant role in the overall memory usage and the practical deployability of SuS. The 2.02 GBytes for 100 concurrent users can be easily handled by modern web servers. The web server hosting the LMS has ample memory and can easily scale to support the more than 1,000 active users in the system. Likewise, the HotCRP service can handle one hundred concurrent users with less than 3 GByte of RAM, which may meet the needs of most conferences. For high-volume websites such as Word-Press, when considering our FMSK improvement, the memory usage will reduce to be relatively the same as the LMS application. But even without further optimization, compared with CLAMP's VM-based approach, each SuS container's memory usage is only half as much as a VM-based Webstack's (64 MB).

Table 4. The average merge time (t_m) and per-container memory saving (m_s) comparison between our FKSM and the UKSM approach across 10 trials with varying container counts. UKSM requires parameter tuning for best performance; the default works better in workloads with < 25 containers.

Concurrent Users					
KSM Approach		10	25	50	100
t_m **(secs)**	Our FKSM	1.24	3.24	8.51	18.56
	UKSM [40]	95.75	88.5	105.25	143.25
m_s **(MiB)**	Our FKSM	8.92	9.84	10.02	10.03
	UKSM [40]	4.98	5.30	5.14	5.00

We compare UKSM with our own Focused Kernel Same-page Merging (FKSM) approach, in which the container manager actively initiates the merging requests and records merge completion time and the average memory saved for each container. In contrast, UKSM constantly runs in the background without a clear merge completion point. Therefore, to make a fair comparison, we record the memory statistics for 10 min and choose the time needed for UKSM to complete 80% of memory savings for its performance statistics. We show these results in Table 4. FKSM focuses on mergeable pages only between pairs of processes, enabling quick merging. In contrast, UKSM makes multiple rounds of local and global samplings and may not discover all merging opportunities (resulting in 48% less memory saving on average). In the shared region of Table 2, we show the FKSM savings, which average around 30%. We examine how web retrievals can lead to unique states to process requests. We found that the memory usage will increase by around 2MB for both page-sharing approaches. Our approach still saves 26% of memory per container. We also found that when ASLR is enabled, the memory saving of FKSM and UKSM is significantly reduced to less than 5%. While the original UKSM paper [40] reports 39% of memory saving for containers, we found that this result is only achievable with a fully duplicated LAMP stack, where most saving is attributed to duplicated MySQL processes. In our settings, multiple containers share a single database with different credentials.

Table 5 shows the median CPU usage (obtained using `mpstat`) for a four-core system with each tested case across 10 trials using the same container configuration as our per-container memory usage experiment with WordPress as the SuS-hosted application. The CPU usage difference between SuS and a shared server appears related to the processes needed to fork and isolate containers. For SuS, each container has a complete process set. For the shared server setup, it only needs to run a single server application and shared PHP worker processes. On the initial load, the 100 concurrent users each cause the accessed PHP scripts to be compiled on SuS, whereas in the shared server, a single compilation suffices due to PHP's OPcache [28]. This likewise explains the closer results for SuS and the shared server models on subsequent loads.

Table 5. The CPU usage (in percent) for both the shared and Single-use Server across 10 trials. For both deployments, the first load on WordPress triggers a bootstrapping process that causes the CPU usage to be higher than the second (and subsequent) page loads. In SuS, the first page load also causes the Container Manager to assign a container to the newly-connected user.

Concurrent Users					
Configuration		10	25	50	100
First Load	Shared	21.01%	23.19%	31.49%	41.33%
	SuS	56.85%	77.02%	85.94%	93.62%
Second Load	Shared	4.36%	10.10%	17.53%	34.83%
	SuS	16.26%	22.40%	27.60%	39.50%

To avoid the PHP bootstrapping and compiling process, as mentioned above, we configured the Container Manager to perform this bootstrapping when creating a SuS container to prime it. Our implementation handles this by sending a pre-provision request to the fresh-started SuS container. We note that such a process can also be configured by the Opcache preload [29] feature but requires specifying the scripts in the correct dependency order to compile. Our request-based approach avoids this requirement.

6.2   Page Retrieval Times

We examine and compare the latency between SuS and Shared server using WordPress, which is known to be a heavyweight application [4]. We generate concurrent client requests using multiple instances of `wget` to get WordPress's main page (each main page access requires 7 different assets files and triggers 22 unique MySQL queries). While website complexity varies in practice, this experiment compares the workload's impact across different server configurations.

Figure 3 shows page load times under different settings. As we mentioned above, one overhead is the multiple script compilation process. Given this, the first two settings for SuS are with and without pre-provision. In addition, we consider a third pool refilling setting where the container manager maintains a pool watcher thread to ensure sufficient available containers.

From Fig. 3, we see that when using pre-provisioned containers, the load times for the SuS and shared server models are similar, from 10 to 50 concurrent users. The SuS model becomes slower at 100 concurrent users. We believe there are two main sources of delay. First, the SuS model requires the CPU to spend extra cycles to context switch between processes, which is saved for shared servers because of sharing worker processes across requests. Second, our Client-to-SuS middlebox must request a new container from the Container Manager for the first request from a new client, adding initial latency and load. For the pool refilling

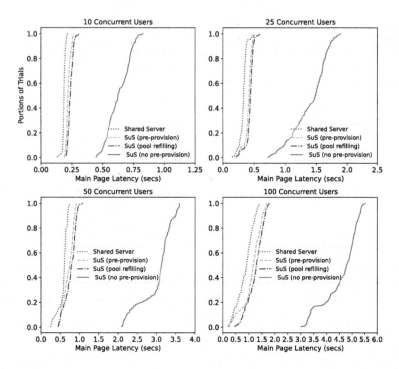

Fig. 3. CDFs of 200 home page load trials in WordPress 5.1.1. with 10, 25, 50, 100 concurrent users in SuS and shared server ("control") scenarios.

setting, we observed only a minor impact on the page load time (characterized by the difference between the second (orange) and third (green) lines in Fig. 3).

In CPU usage and page load time, SuS has results close to a shared server. When memory is not the bottleneck, SuS is able to achieve similar throughput as a shared server. This significantly outperforms the VM-based approach in CLAMP, which had only 42% of the throughput of a shared server [27].

6.3   PHP Profiling Overhead

Our PHP profiling module adds extra runtime procedures to obtain the function's execution context. Then it asynchronously sends the collected traces to a profiling data receiver. The profiling overhead does not affect each asset retrieval, only the request handled by PHP. This experiment measures the overhead by comparing the PHP request's round trip time with and without the module.

Figure 4 indicates that the profiling adds around 10ms delay on individual PHP requests. WordPress makes many function calls (around 25,000 user-defined functions for each PHP request on average). Given this statistic, our profiling module adds less than $1\mu s$ for each function call. The PHP profiling overhead does not need to affect production traffic since analysts can disable this functionality in normal usage and only enable it in *post hoc* analysis in which the

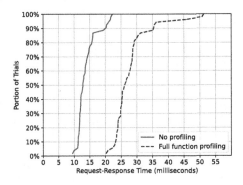

Fig. 4. Profiling affects page load time. Unless applied to production traffic, this logging overhead would occur in *post-hoc* analysis.

profiling module is enabled in a pristine container, and the previously logged HTTP request and back-end resource requests are replayed. Tools like TCPreplay [9] enable such event-based traffic replaying. Accordingly, defenders may choose whether to enable the feature for live traffic or only in incident response. Since our system prunes unrelated interactions, the replay logs may be small.

7   Conclusion

Our work introduces SuS containers that prevent adversaries from exploiting vulnerabilities in front-end Internet servers. These protections require only small code base alterations. Overheads introduced by the containerization approach are limited, with 2GB RAM sufficient for 100 containers and only a 5% increase in CPU consumption. The memory consumption of the approach is practical in some settings. Our FKSM saves 26% of memory for active containers. High-volume servers may benefit from future work in copy-on-write container cloning. This approach captures logs at the middleboxes and execution engine. The approach can allow analysts to reconstruct an incident and localize a vulnerability.

Acknowledgements. This material is based upon work supported by the National Science Foundation under Grant No. 1814402 and 1814234.

References

1. Cybersecurity Help: Privilege escalation in GDPR compliance plugin for wordpress. https://www.cybersecurity-help.cz/vdb/SB2018111101
2. Agache, A., et al.: Firecracker: lightweight virtualization for serverless applications. In: USENIX NSDI (2020)
3. Akamai: web attacks and gaming abuse. State of the Internet **5**(3), 1–30 (2019). https://www.akamai.com/us/en/multimedia/documents/state-of-the-internet/soti-security-web-attacks-and-gaming-abuse-report-2019.pdf

4. Budding, R.J.: Wordpress PHP performance benchmark. https://www.savvii.com/blog/wordpress-php-performance-benchmark-2019/ (2019)
5. Bulekov, A., Jahanshahi, R., Egele, M.: Saphire: sandboxing PHP applications with tailored system call allowlists. In: USENIX Security Symposium (2021)
6. Cheng, R., et al.: Radiatus: a shared-nothing server-side web architecture. In: ACM Symposium on Cloud Computing (2016)
7. Cornelissen, B., Zaidman, A., Holten, D., Moonen, L., van Deursen, A., van Wijk, J.J.: Execution trace analysis through massive sequence and circular bundle views. J. Syst. Softw. **81**(12), 2252–2268 (2008). https://doi.org/10.1016/j.jss.2008.02.068
8. Dunlap, G.W., King, S.T., Cinar, S., Basrai, M.A., Chen, P.M.: Revirt: enabling intrusion analysis through virtual-machine logging and replay. SIGOPS Oper. Syst. Rev. **36**(SI), 211–224 (2003). https://doi.org/10.1145/844128.844148
9. Fred Klassen: tcpreplay-github. https://github.com/appneta/tcpreplay
10. Frei, A., Rennhard, M.: Histogram matrix: log file visualization for anomaly detection. In: IEEE Conference on Availability, Reliability and Security (2008)
11. Garg, A., Mishra, D., Kulkarni, P.: Catalyst: GPU-assisted rapid memory deduplication in virtualization environments. In: ACM SIGPLAN/SIGOPS International Conference on Virtual Execution Environments (2017)
12. Gupta, D., et al.: Difference engine: harnessing memory redundancy in virtual machines. Commun. ACM **53**(10), 85–93 (2010)
13. Hardy, N.: The confused deputy: (or why capabilities might have been invented). SIGOPS OS Rev. **22**(4), 36–38 (1988)
14. Kohler, E.: Hotcrp software. https://github.com/kohler/hotcrp
15. Manco, F., et al.: My VM is Lighter (and Safer) Than Your Container. In: ACM Symposium on Operating Systems Principles (2017)
16. Mehta, A., Elnikety, E., Harvey, K., Garg, D., Druschel, P.: Qapla: policy compliance for database-backed systems. In: USENIX Security (2017)
17. Neuman, B.C., Ts'o, T.: Kerberos: an authentication service for computer networks. IEEE Commun. Mag. **32**(9), 33–38 (1994)
18. Nisi, D., Bianchi, A., Fratantonio, Y.: Exploring Syscall-based semantics reconstruction of android applications. In: USENIX RAID Symposium (2019)
19. NIST: CVE-2009-2762. https://nvd.nist.gov/vuln/detail/CVE-2009-2762
20. NIST: CVE-2012-3578. https://nvd.nist.gov/vuln/detail/CVE-2012-3578
21. NIST: CVE-2019-9879. https://nvd.nist.gov/vuln/detail/CVE-2019-9879
22. NIST: CVE-2019-9880. https://nvd.nist.gov/vuln/detail/CVE-2019-9880
23. NIST: CVE-2020-13693. https://nvd.nist.gov/vuln/detail/CVE-2020-13693
24. NIST: CVE-2021-24182. https://nvd.nist.gov/vuln/detail/CVE-2021-24182
25. NIST: CVE-2021-24183. https://nvd.nist.gov/vuln/detail/CVE-2021-24183
26. Oexman, D.: Changeset for wp-gdpr-compliance. https://plugins.trac.wordpress.org/changeset/1970366/wp-gdpr-compliance (2018)
27. Parno, B., McCune, J.M., Wendlandt, D., Andersen, D.G., Perrig, A.: Clamp: practical prevention of large-scale data leaks. In: IEEE Security and Privacy (2009)
28. PHP Devs.: OPcache. https://www.php.net/manual/en/book.opcache.php
29. PHP Devs.: Preloading manual. https://www.php.net/manual/en/opcache.preloading.php
30. Popov, N.: Extension exposing PHP 7 abstract syntax tree. https://github.com/nikic/php-ast
31. Provos, N., Mavrommatis, P., Rajab, M.A., Monrose, F.: All your iFRAMEs point to us. In: USENIX Security, pp. 1–15. USENIX, USA (2008)
32. Puentes, M.A.: PEGASUS: Powerful, Expressive, Graphical Analyzer for the Single-Use Server. Thesis, Worcester Polytechnic Institute (May 2021)

33. Raoufi, M., Deng, Q., Zhang, Y., Yang, J.: PageCmp: bandwidth efficient page deduplication through in-memory page comparison. In: IEEE Computer Society Annual Symposium on VLSI (2019)
34. Salah, T., Jamal Zemerly, M., Chan Yeob Yeun, Al-Qutayri, M., Al-Hammadi, Y.: The evolution of distributed systems towards microservices architecture. In: IEEE International Conference for Internet Technology and Secured Transactions (2016)
35. Stubbs, J., Moreira, W., Dooley, R.: Distributed systems of microservices using docker and serfnode. In: IEEE International Workshop on Science Gateways (2015)
36. Taylor, C.R.: leveraging software-defined networking and virtualization for a one-to-one client-server model. Master's thesis, WPI (2014)
37. W3Techs: usage statistics and market share of WordPress. https://w3techs.com/technologies/details/cm-wordpress (2020)
38. W3Techs: usage statistics of server-side programming languages for websites. https://w3techs.com/technologies/overview/programming_language (2020)
39. WordPress.org: WordPress. https://www.wordpress.org/ (2003)
40. Xia, N., Tian, C., Luo, Y., Liu, H., Wang, X.: UKSM: swift memory deduplication via hierarchical and adaptive memory region distilling. In: USENIX Conference on File and Storage Technologies (2018)
41. You, L., Li, Y., Guo, F., Xu, Y., Chen, J., Yuan, L.: Leveraging array mapped tries in KSM for lightweight memory deduplication. In: IEEE NAS Conference (2019)

Analysis of Vulnerable Code

Extended Abstract: Towards Reliable and Scalable Linux Kernel CVE Attribution in Automated Static Firmware Analyses

R. Helmke[✉] and J. vom Dorp

Fraunhofer FKIE, Zanderstraße 5, 53177 Bonn, Germany
{rene.helmke,johannes.vom.dorp}@fkie.fraunhofer.de

Abstract. In vulnerability assessments, software component-based CVE attribution is a common method to identify possibly vulnerable systems at scale. However, such version-centric approaches yield high false-positive rates for binary distributed Linux kernels in firmware images. Not filtering included vulnerable components is a reason for unreliable matching, as heterogeneous hardware properties, modularity, and numerous development streams result in a plethora of vendor-customized builds. To make a step towards increased result reliability while retaining scalability of the analysis method, we enrich version-based CVE matching with kernel-specific build data from binary images using automated static firmware analysis. In a case study with 127 router firmware images, we show that in comparison to naive version matching, our approach identifies 68% of all version CVE matches as false-positives and reliably removes them from the result set. For 12% of all matches it provides additional evidence of issue applicability.

1 Introduction

Safety, security, and privacy threats arise alongside embedded system markets. Growing device numbers inflate attack surfaces, raising impact and scope of newly found software vulnerabilities in domains pivotal to society [8]. Thus, it is important to maintain the software security of these systems.

Embedded devices commonly make use of Embedded Linux[1] as host operating system for their firmware. Using open source components instead of developing custom solutions generally provides a solid security foundation; though the Linux kernel specifically has been attributed over 2,900 Common Vulnerabilities and Exposures (CVE)s as of 2022. Attesting the security of a Linux-based firmware thus includes checking which CVEs concern the specific kernel in-use.

While reproducible exploitation of a CVE would be optimal, various challenges [10,13] exist that make a comprehensive reproduction on a device unobtainable. First, not all CVEs have a known POC exploit to test against. Second,

---

[1] https://www.embedded.com/wp-content/uploads/2019/11/EETimes_Embedded_2019_Emb\discretionary-edded_Markets_Study.pdf.

© The Author(s), under exclusive license to Springer Nature Switzerland AG 2023
D. Gruss et al. (Eds.): DIMVA 2023, LNCS 13959, pp. 201–210, 2023.
https://doi.org/10.1007/978-3-031-35504-2_10

exploitation requires either a running device or an emulated firmware. The former is not generally attainable for large-scale analysis. The latter is hindered my various challenges specific to firmware emulation [6,13].

Static analysis serves heuristics to find *imperfect* proof for CVE attribution. Yet, many approaches do not scale well as they require considerable manual work and deep knowledge of each CVE [10]. Parts are automatable but needed data may be unavailable or incorrect in repositories [1,11]. Also, automation becomes increasingly challenging considering proprietary formats, obfuscation, compiler optimizations, and symbol stripping [3,10].

In lack of better methods, firmware analysis tools [4,9] and large-scale studies [2,12,14] commonly attribute vulnerabilities by matching versions against CVE databases. One such study [12] used version matching on the Linux kernel as part of an empirical study on home router security. Due to custom build configurations, implying that each kernel includes only a subset of all possible vulnerabilities, this method is exceedingly unreliable. To improve the reliability of version-based Linux CVE attribution in large-scale scenarios, we enrich such naive matching with kernel-specific configuration data, collected through automated static firmware analysis. Hereby, we reduce the set of false-positive matches requiring further manual verification. In the following, we provide:

1. A description of our methodology for Linux CVE attribution, based solely on binary kernel representations.
2. A case study in which we compare our approach with naive version-based CVE matching using the 2020 Home Router Security Report [12] dataset.
3. An open source proof of concept implementation of the methodology[2].

2   Background & Related Work

Automated vulnerability detection is approached using various methods such as code similarity and patch analysis [5], fuzzing [7], and various emulation-based methods [13]. Large-scale detection of known vulnerabilities requires sound ground truth. Thus, we focus our discussion on sound vulnerability information and previous research on discovering known vulnerabilities on binary code.

Sound Data as Foundation for known Vulnerability Detection. Correct and detailed information on known vulnerabilities [10] is essential for effective automated detection methods. The community-driven CVE catalog[3] offers a de facto standard for vulnerability identification but comes with limitations due to errors in Common Platform Enumeration (CPE) assignments [1], missing or hard to obtain data [3] and inconsistent references to patches [11]. Additionally, for CVEs affecting closed source projects, issuers will not share technical details on fixes in public. In this work, we leverage upon the observation that the summary of most Linux kernel CVEs includes a file reference to mark which kernel part is affected.

---

[2] https://github.com/fkie-cad/cve-attribution-s2.
[3] https://www.cve.org/.

In consequence, most research comes with small custom datasets of selected CVEs that their proposed techniques can ingest for evaluation [10]. The necessary investigation, data aggregation, and technical bug knowledge, limits the applicability of previously -scale scenarios.

Static Vulnerability Detection in Large-Scale Firmware Analyses. In 2014, Costin et al. [2] executed a quantitative study on embedded device security by analyzing 32.000 firmware images. They attributed CVEs based on software version number and then reported unsolved challenges in result verification, as not only CVE data is incomplete, but vendors may also custom-patch files.

Cross-architecture code similarity methods (e.g. FirmUp [3]) have drastically improved and may be used as imprecise measure for verification in this case. However, acquiring and processing patches for thousands of CVEs to bootstrap code similarity methods deems infeasible based on the imprecise CVE repositories.

Zhao et al. [14] develop FirmSec, a large-scale static analysis pipeline for IoT devices. The approach extracts syntactical and control-flow graph features and, thus, provide an alternative for signature-based version detection. However, the applicability issue introduced by vendor-specific build configurations, as in the Linux kernel, is not considered.

The authors of [12] assess and compare the state of firmware security of 127 home routers in similar aspects as [2], using the automated Firmware Analysis and Comparison Tool (FACT) [4]. Identified Linux kernel versions are matched against the National Vulnerability Database (NVD)[4] to calculate how many critical CVEs affect the kernel of each firmware. In this study, the stated issues with CVE database information and kernel modularity lead to high false-positive rates, incurring high manual verification efforts.

There is few work that specifically studies high false-positive CVE attribution rates caused by the Linux kernel's modularity. With version-based CVE attribution being a common method, we identify false-positive reduction in static kernel CVE attribution as a research gap.

3   Methodology

This section describes our proposed methodology to enrich the version-based Linux kernel CVE attribution process with build-specific annotations. We show an automated static analysis pipeline that finds and extracts kernel configurations, dry builds the found kernel version, and filters CVEs based on affected version and build log-included files.

Figure 1 provides an overview of our methodology. We establish a two-stage process: In the first and left-hand stage, we unpack, analyze, and annotate each file of an ingress firmware image. Gathered information includes Linux kernel version, Instruction Set Architecture (ISA), and kernel build configuration. In the second and right-hand stage, we leverage upon said data to perform the actual CVE attribution and filtering step. Yellow boxes in Fig. 1 mark components this paper contributes.

---

[4] https://nvd.nist.gov/.

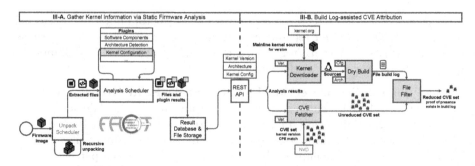

Fig. 1. Two-staged static analysis pipeline to (I) gather kernel information and (II) attribute kernel CVEs accurately.

In the two following Subsects. 3.1 and 3.2, we provide detailed technical insights on each stage and step.

3.1 Gather Kernel Information via Static Firmware Analysis

For stage one, we apply and enhance the open source firmware analysis tool FACT [4]. FACT provides automated firmware analysis capabilities including recursive extraction, kernel version, and ISA detection. In the following, we describe all steps that are of importance for the proposed attribution methodology.

Starting with an arbitrary Linux firmware image, we first use FACT internals to recursively extract all components necessary for analysis, including the kernel. Next, we identify the ISA and kernel version. The **Analysis Scheduler** achieves this by running a selected set of analyses on each extracted object. The software version detection uses YARA rules and the ISA detection leverages ELF header information, detected kernel configurations, and device trees.

We contribute the **Kernel Configuration** plugin, which detects and extracts Linux kernel build configurations in firmware images. It is pivotal to the succeeding dry build pipeline step, as it determines components included in kernel builds. In firmware, kernel configurations may be present as plain text or in binary form. Detection of plain text configurations is straight forward due to the distinctive key-value structure and well-known directive keywords. These can be used for pattern matching. If the CONFIG_IKCONFIG directive is enabled (Y) during build, the kernel configuration gets embedded into the binary kernel image. This embedding might be an inline string or a binary compressed representation using common algorithms like LZMA or DEFLATE. If it is set to M, the configuration is outsourced to a kernel module. Thus, if the file is either a kernel image or module, our plugin searches for an embedded magic word that precedes the kernel configuration data. The plugin tests for all variations and extracts, and if necessary, decompresses the configuration.

3.2 Build Log-Assisted CVE Attribution

The build log-assisted CVE attribution is the second stage of our proposed analysis pipeline in Fig. 1. Here, we first use FACT's **REST API** to consolidate the kernel version, kernel build configuration, and detected target ISA.

Then, our contributed **Kernel Downloader** fetches mainline version sources from `kernel.org`. We emphasize that our assumption of unaltered mainline kernels in firmware images is likely false because vendors may custom-patch their kernels. However, we observe that modified kernel code is not accessible in scale, regardless of the Linux kernel's GNU General Public License (GPL) that dictates vendors to publish modified open source code. For example, some vendors complicate distribution by implementing manual request procedures for each device, firmware, and version[5].

Dry Build is the next step in Fig. 1. We set the target ISA and install the extracted kernel build configuration in the downloaded kernel source project. Then, we execute a compilation dry run, which does not compile the kernel but prints each compilation recipe instead. This approach has the advantages of low computational overhead and no requirement for a cross-compilation toolchain. With this step, we gather a list of source files from the build log, which our pipeline *witnesses* to be included in the kernel build.

The **CVE Fetcher** executes simultaneously. We query the NVD dataset for all Linux kernel CVEs and filter out all records that do not refer CPEs stating the extracted Linux kernel version to be vulnerable. The result of this stage is identical to naive version-based attribution.

The **File Filter** step combines the outputs of CVE Fetcher and Dry Build: Based on the observation that Linux kernel CVEs summaries usually state the affected source files, we improve on the version-based attribution by removing every CVE that does not reference an affected file we witnessed in the build log.

4   Case Study

We perform a case study to evaluate the reliability of our enriched version-based Linux kernel CVE attribution in large-scale static analyses. For this purpose, we let our pipeline analyze the Home Router Security Report 2020 [12] corpus, which vendors reported to yield high false-positive rates using version-based CVE matching. We raise two research questions:

R1 Our methodology requires access to specific information in firmware samples and CVE repositories. How many samples and CVEs fulfill these modalities? *How applicable is our approach in a real-world scenario?*

R2 With version-based CVE matching as baseline, *what impact has the methodology on result reliability?*

In the following subsections, we first provide detailed information on our experiment and used dataset (4.1). Then, we present the results and analyze them within the context of both stated research questions (4.2 and 4.3, respectively).

4.1   Experiment & Firmware Corpus

Experiment Execution. We deploy our analysis pipeline on a x86_64 desktop system, running Ubuntu 20.04.4 LTS. FACT v4.0 (commit `38df4883`) is used

---

[5] https://www.zyxel.com/form/gpl_oss_software_notice.shtml.

in the first pipeline stage to detect CPU architecture, identify the kernel, and extract the kernel configuration. The second pipeline stage executes on the same machine based on a snapshot of the NVD – taken on 2022–08-30. The snapshot has records for 2,910 Linux kernel CVEs attributable through CPE. For each component in our system, we collect details on ingress and egress data, including plugin results, version-based CVE matches, and filtering decisions.

Firmware Corpus. The analyzed Home Router Security Report [12] corpus is publicly documented[6], and consists of firmware from 127 recent home routers. Devices of seven vendors are included: ASUS, AVM, D-Link, Linksys, Netgear, TP-Link, and Zyxel. Samples were scraped on 2020–03-27.

Across all 127 samples, 121 binary distributed Linux kernels from v2.4.20 to v4.4.60 are included. The most common major version is 3.x with 49 kernels, while 44 kernels have version 2.6. 11 firmware images are not analyzable due to failed operating system detection or unpacking errors. Note that firmware can contain multiple kernels, e.g., embedded devices may consist of subcomponents running their own systems.

All identified CPU architecture belong to the MIPS and ARM ISAs, with a majority having a word length of 32-bit. The ISA is unknown for 24 samples. Further corpus insights can be found in [12].

While we acknowledge the missing size and device class diversity of the dataset compared to studies like [2,6], we argue that the dataset is of sufficient size to demonstrate applicability, as it covers Linux kernels from three major releases, widely spread ISAs, and devices of multiple vendors. We also choose it to investigate the reliability of matches reported in [12] using naive version-based Linux kernel CVE attribution.

4.2   R1 Analysis – Applicability in Real-World Scenarios

We identify two methodological requirements that must be fulfilled for each firmware and Linux CVE for our approach to succeed:

S1 FACT firmware extraction and all plugin analyses of stage one must succeed to consolidate the kernel version, ISA, and kernel configuration.
S2 CVE descriptions must reference affected files to filter vulnerable components not included in the kernel build.

Using the firmware corpus, we evaluate egress and ingress data of each step in the proposed pipeline with regards to these requirements. Table 1 shows the results. Highlighted rows designate effective requirement fulfillment rates over all analyzed firmware images and Linux kernel CVEs.

For requirement S1, FACT successfully extracted 116 out of 127 firmware images. It then identified both kernel binary and kernel version for all 116 extracted images. The ISA was successfully identified in 103 images. However, our Kernel Configuration plugin finds build information in only 44 samples. Thus 34.64% of all analyzed firmware samples fulfill requirement S1. This rate is explainable

---

[6] https://github.com/fkie-cad/embedded-evaluation-corpus/blob/master/2020/FKIE-HRS-2020.md.

Table 1. Method Applicability Analysis for the Firmware Corpus

S1 Requirement (FACT Analysis Success)		
	FW Matches [#]	Fulfilled $\left[\frac{\text{FW Matches}}{\text{FWs Total}}\right]$
Extraction	116	0.9133
Kernel Version	116	0.9133
Architecture Detection	103	0.8110
Kernel Configuration	44	0.3464
S2 Requirement (File Reference in Linux Kernel CVE)		
	CVE Matches	Fulfilled $\left[\frac{\text{CVE Matches}}{\text{CVEs Total}}\right]$
Full Path Reference	1743	0.5990
File Only Reference	129	0.0443
No Reference	1038	0.3567

considering that a) `IKCONFIG` must be explicitly enabled to embed kernel configurations into binary representations and b) it is common practice for vendors to strip unnecessary information for memory saving and obfuscation purposes.

For requirement S2 (affected files must be referenced in Linux kernel CVE descriptions), data analysis over all Linux CVEs inside the NVD yields three different categories: Files are either referenced as **Full Path** relative to the kernel's source tree, or the reference is **File Only** (location in the source tree is unknown), or **No Reference** exists at all. Table 1 distributes all 2,910 Linux kernel CVEs across these classes, showing that the proposed approach is applicable to 1,872 (64.33%) kernel CVEs. For CVEs with no included file reference, the approach falls back to version-based CVE matching and, thus, cannot add value to result reliability.

4.3   R2 Analysis – Impact on CVE Attribution Result Reliability

We approach research question R2 by analyzing the attribution results of all 44 firmware images our methodology is applicable to (cf., Sect. 4.2). Subject samples

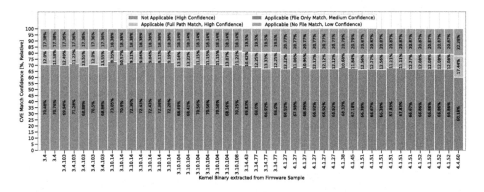

Fig. 2. Filter verdict distribution of our pipeline relative to the baseline CVE attribution results for each of the 44 analyzed kernels.

include kernels ranging from v3.4.0 to v4.4.60. At the time of this evaluation, none of these are still actively maintained by the mainline kernel team.

The baseline method attributes a median of 1,196 CVEs per firmware image, which is roughly 40% of *all* Linux kernel CVEs present in the NVD. A possible explanation lies within unsound and/or unmaintained CVE records in the NVD [1,11]. Based on the results we present in the following paragraphs, there is reason to assume that the baseline yields exceedingly high false-positive rates.

Version-based CVE attribution is an intermediate result of our methodology (cf., Sect. 3). To estimate the impact our pipeline has on result reliability, we consolidate all decisions of the build log-assisted filtering to classify them into four categories of verdict confidence:

- **Applicable (High)** – CVE references affected files and full file path is witnessed in build log.
- **Not Applicable (High)** – CVE references affected files but none of them is present in the build log.
- **Applicable (Medium)** – CVE references affected files but does not state full file paths. A file was matched and seen in the build log, but ambiguity exists due to duplicate names in the source tree.
- **Applicable (Low)** – No file references, we cannot decide on applicability and fall back to version-based matching.

The idea is to map persuasiveness of additional evidence the pipeline gathers within a trial: File matches are witnesses for CVE applicability, but not every match is equally credible.

Figure 2 shows the filter verdict distribution of our pipeline relative to the baseline CVE attribution results for each analyzed kernel. Versions are ordered from oldest (left) to newest (right). Note that a single kernel was found in each one of the 44 analyzed samples. Thus, each entry on the horizontal axis represents a unique firmware. All distribution values are medians across all samples.

The proposed Linux kernel CVE attribution methodology made a medium to high confidence decision for 80.6% of all version-based matches. The portion of high confidence applicable CVE matches is 12.04%. Relative path matches yielding medium confidence applicability are negligible with 0.19%. As indicated by the bottom bars belonging to the class of Not Applicable (High), our pipeline attributes 68.37% of all version-based CVE matches as *false-positives* and filters them out of the result set. Out of the median 1,196 matches per firmware, we reduce the set of potentially applicable CVEs to roughly 378. Thus, we significantly reduce the result set of potentially applicable CVEs requiring manual verification by analysts and vendors. The portion of low confidence applicability due to missing file references is 19.4%. Unfortunately, our methodology does not generate added value for this subset.

5 Limitations

We identify methodological shortcomings in three dimensions: applicability, sound ground truth, and functionality.

In terms of applicability, our Linux kernel CVE attribution pipeline is bound to FACT's static analysis success. If the kernel version, ISA, and build configuration remain unknown, our method cannot identify possibly included components for reliable CVE filtering. Yet, the case study in Sect. 4 shows that there is still a considerable amount of firmware fulfilling all requirements.

As for sound ground truth, reliable and true-positive CVE attribution is limited by the quality of its underlying dataset. Unsound Linux kernel CVE records that reference unaffected versions or source files can introduce false matches in our proposed method. Our assumption of vendors using mainline kernels is another limiting factor that affects reliability, but a methodical necessity due to missing insider information. Vendors may cherry-pick patches or introduce custom fixes, which are not detectable by our approach. While some of the modifications might be obtainable through GPL portals, we identify the issue of scalable accessibility. Another limitation comes from the file-based filtering. Kernel builds can in- or exclude only parts of a file based on configuration options. This can lead to exclusion of vulnerable code, while the *affected* file still appears in the build log.

Regarding functional limits, we stress the inherent limitations of static analysis. It may use heuristics to find indicators of bug presence but can hardly serve definitive proof – which usually requires triggering the bug during runtime.

Finally, the conducted case study is limited in its validity, as the used corpus lacks device class heterogeneity.

6   Conclusion and Future Work

In this paper, we present a method to improve result reliability of version-based CVE matching for the special case of binary Linux kernels in large-scale static firmware analyses. This is achieved by enriching naive version-based CVE matching with a static attribution pipeline that detects kernel configurations and ISAs in firmware images. We reconstruct the kernel build process and infer included source files. Combined with kernel CVE information, where affected files are explicitly stated, this can be used to remove most false-positive CVE annotations.

The case study shows that, with the limitations discussed in Sect. 5 in mind, our method is scalable and moderately applicable: For 34.64% of firmware images, the technical requirements are fulfilled and about 65% of all Linux kernel CVEs reference affected files in their description. Compared to naive version-based matching, the method generates a high-confidence filter verdict for 80.6% of all attributed CVEs. Specifically, 68.37% of attributed CVEs are discarded as false-positives. We contributed stage one of our pipeline to the publicly available FACT [4] and published the scripts for stage two on GitHub (https://github. com/fkie-cad/cve-attribution-s2).

In future work, we want to address the reliance on inline kernel configuration that leads to the moderate applicability by researching alternative options to infer configuration from binary kernels. Also, the fine-granular commit-based version tracking as offered by `linuxkernelcves.com` is a promising alternative data source for initial version-based attribution. Furthermore, partial file compilation and custom backports should be addressed. Finally, we plan a large-scale evaluation addressing missing device class heterogeneity, like [2,6].

References

1. Benthin Sanguino, L.A., Uetz, R.: Software Vulnerability Analysis Using CPE and CVE. ArXiv (2017). https://doi.org/10.48550/arXiv.1705.05347
2. Costin, A., Zaddach, J., Francillon, A., Balzarotti, D.: A Large-Scale Analysis of the Security of Embedded Firmwares. In: 23rd USENIX Conference on Security Symposium (SEC '14). USENIX Association, San Diego, USA (2014)
3. David, Y., Partush, N., Yahav, E.: FirmUp: precise Static Detection of Common Vulnerabilities in Firmware. In: 23rd International Conference on Architectural Support for Programming Languages and Operating Systems (ASPLOS '18). ACM, Williamsburg, USA (2018)
4. Fraunhofer FKIE: FACT - Firmware Analysis and Comparison Tool. https://github.com/fkie-cad/FACT_core
5. Haq, I.U., Caballero, J.: A Survey of Binary Code Similarity. ACM Comput. Surv. **54**(3), 01–38 (2021)
6. Kim, M., Kim, D., Kim, E., Kim, S., Jang, Y., Kim, Y.: FirmAE: towards Large-Scale Emulation of IoT Firmware for Dynamic Analysis. In: 2020 Annual Computer Security Applications Conference (ACSAC '20). ACM, Austin, USA (2020)
7. Manès, V.J., et al.: The Art, Science, and Engineering of Fuzzing: a Survey. IEEE Trans. Softw. Eng. 47(11), 2312–2331 (2021)
8. Neshenko, N., Bou-Harb, E., Crichigno, J., Kaddoum, G., Ghani, N.: Demystifying IoT Security: an exhaustive survey on iot vulnerabilities and a first empirical look on internet-scale IoT exploitations. IEEE Commun. Surv. Tutorials 21(3) (2019)
9. ONEKEY GmbH: ONEKEY Automated Firmware Analysis Platform. Accessed 5 Sep 2022. https://onekey.com/
10. Qasem, A., Shirani, P., Debbabi, M., Wang, L., Lebel, B., Agba, B.L.: Automatic Vulnerability Detection in Embedded Devices and Firmware: survey and layered taxonomies. ACM Comput. Surv. **54**(2), 1–42 (2021)
11. Tan, X., et al.: Locating the Security Patches for Disclosed OSS Vulnerabilities with Vulnerability-Commit Correlation Ranking. In: 2021 ACM SIGSAC Conference on Computer and Communications Security (CCS '21). ACM, Virtual, Republic of Korea (2021)
12. Weidenbach, P., Vom Dorp, J.: Home Router Security Report 2020. Fraunhofer Institute for Communication, Information Processing and Ergonomics (FKIE), Tech. Rep. (2020). https://www.fkie.fraunhofer.de/en/press-releases/Home-Router.html
13. Wright, C., Moeglein, W.A., Bagchi, S., Kulkarni, M., Clements, A.A.: Challenges in Firmware Re-Hosting, Emulation, and Analysis. ACM Comput. Surv. **54**(1), 1–36 (2021)
14. Zhao, B., et al.: A Large-Scale Empirical Analysis of the Vulnerabilities Introduced by Third-Party Components in IoT Firmware. In: 31st ACM SIGSOFT International Symposium on Software Testing and Analysis (ISSTA '22). ACM, South Korea (2022)

Divak: Non-invasive Characterization of Out-of-Bounds Write Vulnerabilities

Linus Hafkemeyer[1], Jerre Starink[2], and Andrea Continella[2(✉)]

[1] TU Delft, Delft, The Netherlands
linus.hafkemeyer@pm.me
[2] University of Twente, Enschede, The Netherlands
{j.a.l.starink,a.continella}@utwente.nl

Abstract. Despite the high level of automation that fuzzing has brought into the vulnerability research process, the assessment of a discovered vulnerability's implications mostly requires human expertise and intuition. A promising approach to reduce such a manual effort is the automatic extraction of vulnerability characteristics that provide vital clues for exploitability. In this work, we focus on out-of-bounds write vulnerabilities and investigate how to automatically distill the set of source code-level objects affected by such unintended writes. As this poses unique challenges with regard to the invasiveness of the analysis methods, we propose a novel approach that enables monitoring a compiled program for spatial memory safety violations *without* the need for heavy instrumentation. We implement DIVAK, a prototype of our design, and we evaluate it on both benchmarks and real-world vulnerabilities, showing that its detection and characterization capabilities outperform instrumentation-based tools in several scenarios, at the cost of an increased overhead.

Keywords: Vulnerability Analysis · Out-of-bounds Writes

1 Introduction

Out-of-bounds (OOB) writes [1] are still regarded as one of the most dangerous types of software vulnerabilities [22]. Over the years, a vast number of defenses against OOB writes and other memory corruption bugs have been proposed. Preventive approaches such as the deprecation of unsafe functions [16] and memory-safe languages [12,19] reduce the risk but cannot solve the problem without full adoption. Mitigation approaches introduced into operating systems and compilers complicate or prevent exploitation [9,27], but attackers continue to find ways for evading them. Detection approaches based on static and dynamic analysis have been hugely successful [11]. However, only a relatively small share of discovered bugs has relevant security implications. Thus, further investigation into the severity of discovered bugs and prioritization for patching is essential.

As triaging, root cause analysis, and patching of discovered bugs are usually conducted manually by humans, this is often an expensive and time-consuming process, causing potentially severe vulnerabilities to remain unpatched for a long

© The Author(s), under exclusive license to Springer Nature Switzerland AG 2023
D. Gruss et al. (Eds.): DIMVA 2023, LNCS 13959, pp. 211–232, 2023.
https://doi.org/10.1007/978-3-031-35504-2_11

time. Therefore, further automating the process that follows the discovery of a bug can substantially decrease the time required to develop a patch. Unfortunately, this process is often an intricate task that largely relies on human expertise and intuition, making full automation difficult.

Recent research [32] has shown that *partial* automation is a promising alternative to approaches based on fully Automatic Exploit Generation (AEG) [3,5,14,15,40]. By automatically distilling characteristics of a vulnerability that are decisive for its exploitability, experts can base their assessment of the bug's security implications on a concise high-level summary, accelerating the triaging process. Besides, AEG based on human-interpretable vulnerability characteristics can make intermediate results substantially more helpful for manual analysis.

We study how to automatically distill such characteristics of OOB writes from programs. In contrast to the state-of-the-art, we aim to extract capabilities that are truthful to the form in which the program under test is deployed in practice, thus revealing vulnerability characteristics that are relevant *not only in laboratory or debugging settings* but for the program's *real-world usage*. We realize this by developing a system that takes a program under test together with an input that is suspected of causing an OOB write and dynamically performs fine-grained monitoring for OOB writes, mapping affected memory regions to the corresponding source code-level entities and reporting the results in a concise and easily interpretable form. Our system is meant to assist security analysts in triaging potential OOB write vulnerabilities, automating their identification and capability extraction phase.

Designing such a system comes with three main challenges: (1) Many approaches are invasive due to heavy instrumentation, modifying the memory layout and runtime behavior of the program and thus making insights inapplicable to the original program; (2) The compilation to machine code causes large parts of source code semantics to be lost, including entities like variables and data types, as well as information on which entity a specific write to memory is intended to modify according to the program semantics. However, this information is vital for detecting OOB writes, and is essential for achieving easy interpretability of the results; (3) Modern compilers perform optimizations during compilation to increase the program's efficiency, which often causes substantial modifications to the program's machine code-level structure and memory layout.

To address such challenges, we propose a new approach for the dynamic characterization of OOB write vulnerabilities in C programs. Contrary to existing works, our approach does not modify the program through instrumentation and, as such, is entirely *non-invasive*. Instead, we provide a conceptual framework for making source code-level semantic information available to our low-level OOB write-checking technique. As the issue of invasiveness predominantly concerns the stack and global sections, we only focus on the characterization of OOB write vulnerabilities within these regions of programs written in C, compiled with Clang for Linux on AMD64 platforms, and leave out heap-based OOB writes from our analysis. We implement our approach in a system named DIVAK, which achieves a detection rate of 89% on the RIPE benchmark [39], compared to the 70% and 34% achieved by the instrumentation-based current state-of-the-art approaches ASan and SoftBound—at the cost of an increased execution time overhead, along with a slightly increased chance of false positives. Ultimately, DIVAK precisely highlights the source code-level objects affected by OOB writes, assisting humans in triaging potential vulnerabilities.

```
1    struct userEntry {
2      char username[32];
3      int id;
4      bool isAdmin;
5    };
```

```
6    void login(struct userEntry* userPtr) {
7      struct userEntry user = *userPtr;
8      char realPw[32], tryPw[32]; bool pwOk;
9      getUserPassword(&user, realPw);
10     fgets(tryPw, 32, stdin);
11     if (pwOk = !strcmp(realPw, tryPw))
12       setUserLoggedIn(&user);
13   }
```

Fig. 1. Motivating example.

Contributions. We make the following contributions:

- We introduce a technique for low-level bounds-checking by leveraging the intermediate program representation during compilation, overcoming the lack of source code-level semantic information.
- We design a non-invasive OOB write characterization approach able to triage spatial memory safety violations on the stack and in the global sections.
- We implement DIVAK and we evaluate it on artificial benchmarks and real-world vulnerabilities, showing its advantages over state-of-the-art tools.

We make our dataset and code available: https://github.com/utwente-scs/divak

2 Motivation

No existing tool for detecting OOB writes is suitable for characterizing their capabilities and identifying their source code-level consequences. In fact, for our scenario, i.e., triaging potential vulnerabilities in real conditions, all publicly available approaches suffer from one or more of the following limitations.

Inability to Detect Intra-object OOB Writes. Many approaches cannot detect intra-object OOB writes within composite objects such as structures [8, 20,29,31,37,41]. However, intra-object OOB writes are well capable of inducing security issues and need proper triaging, as is illustrated in Fig. 1. Here, an overflow of username can corrupt the isAdmin flag, enabling a non-control data attack. Thus, their inclusion in a vulnerability's capability profile is critical.

Required Hardware Support. Some approaches [8,26,30] rely on extensions to the ISA of the CPU to perform OOB write detection. While such ISA extensions are available for SPARC [30], ARM64 [8], and some historic Intel AMD64 CPUs [26], they are not included in the ISA of any recent AMD64 CPUs.

Invasive Modification of the Program. Existing approaches significantly affect the program's memory layout due to their instrumentation. Such modifications fall into the following categories: (1) Introduction of poisoned red zones around memory objects; (2) Introduction of new memory regions to store metadata, e.g., as shadow memory; (3) Direct and indirect modification of stack frames caused by storing metadata and performing checks. Consider the snippet shown in Fig. 1 and its stack layout as implemented by ASan [29] and SoftBound [23] in Fig. 2. We can clearly see both solutions heavily modify the stack frame layout. This, accompanied with the extra register spilling introduced by the checking logic as well as compiler optimizations on the instrumented program, makes reliably identifying and triaging the memory objects affected by OOB writes within the non-instrumented program challenging.

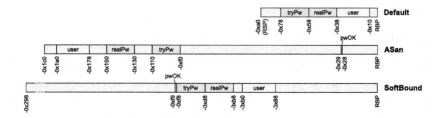

Fig. 2. Stack layouts of the function in Fig. 1 (Default, ASan, and SoftBound).

Table 1. Our approach vs. existing memory sanitizers.

	ASan [29]	HWAsan [8]	Memcheck [31]	SGcheck [37]	SoftBound [23]	Delta Ptrs [20]	Intel MPX [26]	BinArmor [33]	PAriCheck [41]	Our approach
OOB writes detection in globals	✓	✓	×	✓	✓	✓	✓	✓	✓	✓
OOB writes detection on stack	✓	✓	×	✓	✓	✓	✓	✓	✓	✓
OOB writes detection on heap	✓	✓	✓	×	✓	✓	✓	✓	✓	×
Strong spatial guarantees*	×	×	×	×	✓	×	✓	×	×	✓
Intra-object OOB write detection	×	×	×	×	(✓)	×	✓	✓	×	✓
Instrumentation type	CTI	CTI	DBI	DBI	CTI	CTI	CTI	SBI	CTI	n.a.
No need for hardware support	✓	×	✓	✓	✓	✓	×	✓	✓	✓
No mem. layout modification	×	×	✓	✓	×	×	×	×	×	✓
Compatible with external code	✓	×	✓	✓	×	✓	✓(?)	✓(?)	✓	✓
Detection approach	TW	PB	TW	HR	PB	PB	PB	PB	OB	PB
Performance overhead	L	L	H	?	L	L	M	M	L	H
Memory overhead	H	L	?	?	L	L	M	?	L	H

*: Ability to non-probabilistically detect non-continuous and underflowing OOB writes.
TW: tripwire, OB: object-based, HR: heuristics, PB: pointer-based
L: 0x-1x overhead, M: 1x-3x overhead, H: 3x+ overhead, ?: no data

Instability After Out-of-Bounds Write. Most tools terminate the program's execution after detecting one OOB write. While this behavior can often be circumvented, it might cause the program to follow invalid execution paths for the non-instrumented program. Consider a detection approach that uses the stack to store metadata such as bounds information. An OOB write that overwrites this metadata may result in false positives and negatives or let the program crash. As we wish to reliably triage OOB writes during the execution of a program, this limitation makes most existing approaches unusable for our case.

3   Divak: Design

To overcome the challenges described in Sect. 2, we design DIVAK, a pointer-based OOB write detection approach that characterizes spatial memory violations on the stack and in the global sections in a deterministic fashion. By not interfering with the compiled machine code, DIVAK is entirely non-invasive and does not rely on special hardware support. As such, differently from existing work (Table 1), any insights about the effects of OOB write vulnerabilities in the examined binary also hold when the program is not monitored by our approach.

Our approach categorizes memory objects for bounds-checking. For *composite objects* like structs and arrays of structs, we consider the inner structure for the detection of *intra-object* OOB writes. Any other object is instead a *unitary object*—a homogeneous chunk of memory for which we disregard inner structure.

We focus on detecting OOB writes occurring in the static sections `.data` and `.bss`, as well as on the stack. We disregard heap-based OOB writes, as their characteristics are not necessarily distorted by instrumentation-based detection approaches—e.g., SoftBound neither allocates heap memory nor affects the allocator. Thus, DIVAK's main novelty, i.e., its non-invasiveness, is not essential in this scenario. Nevertheless, DIVAK could be extended to support heap-based OOB writes with little engineering effort, and it is compatible with existing tools that target the heap [13,29]. Besides, we assume that target programs are compiled without frame pointer omission and tail call optimization. Finally, as the majority of memory-modifying instructions that can potentially cause OOB writes are those of the `mov` family, we focus our bounds-checking on this family.

3.1   Approach Overview

Our approach takes as input the source code of a program and a proof-of-concept (PoC) input suspected of causing an OOB write and it outputs information about the effects of all discovered OOB writes on source code-level objects. At a high level, we perform three phases: preliminary analysis, static analysis, and dynamic analysis. In the first phase, we instrument the compilation phase of the target program to passively collect information about debug symbols, data structures, and write operations. This information enables us to map properties of source code to compiled code, which allows us to later pinpoint the specific source code-level objects affected by OOB writes. Besides, it also allows us to handle the loss of semantics caused by the compilation process, which is critical to characterize OOB writes *without* requiring invasive modifications of the program.

In the second phase, leveraging the collected information at compile time, we statically analyze the target binary to identify variables and parameters stored on the stack or in the globals, determine their sizes, identify pointer-creating instructions, and determine the destination objects of write operations. Besides, in this phase, we determine the internal structure of composite types such as struct, which is essential to detect intra-object OOB writes.

In the third phase, we dynamically analyze the target program by using the findings obtained through static analysis, taint pointers, and identify write operations that have an effect beyond the boundaries of the intended destination objects (IDOs). Here, we map our results back to the source code domain using the information we collect in the previous phases. Finally, because we do not alter the state or memory layout of the program at run-time, our approach guarantees that, by design, execution continues reliably after detecting an OOB write.

Although the goals of preliminary and static analysis could theoretically be achieved by modifying the compiler, this would come with several drawbacks: (1) Heavy modifications of highly complex code at multiple compilation stages with little documentation; (2) Potentially altered binaries due to modifications, including in production builds; (3) Incompatibility with custom or new optimization passes. Therefore, we opted for the more portable hybrid approach.

Our design for detecting OOB writes relies on the identification and special treatment of the following types of machine code instructions.

Independent Writes. For independent writes in the machine code, the dominant component from which the destination address is computed in the operand is either given by an immediate value or a stack frame boundary register (rbp or rsp). This has two important implications. First, independent writes can only write to global objects or within the stack frame of the containing function. Second, their intended destination object does not change at runtime. An example of an independent write is the instruction mov [rsp + rax], cl, which may access an array on the stack. Here, rsp constitutes the dominant component of the address calculation as its value will be substantially larger than the value in rax. Now, rsp being a stack boundary register makes this an independent write.

Dependent Writes. For dependent writes, the dominant component used to compute the destination address is given by a general-purpose register as opposed to a stack frame boundary register. Thus, dependent writes rely on a previous instruction for determining the pointer used as the basis of the address computation. This requires detaching the logic for determining the intended destination object from the logic for checking the legality of the write. An example of a dependent write is the instruction mov [rcx + rax*8], rdx, which may access an array based at the address specified by rcx. Here, the dominant component is given by a general-purpose register, making this a dependent write.

Pointer-creating Instructions (PCIs). To facilitate bounds-checking of dependent writes, it is essential to taint pointers with their intended pointee object as early as possible by identifying the instructions at which pointers are created.

Bounds-narrowing Instructions (BNIs). Children of composite objects are often accessed by offsetting a pointer to the object. To detect intra-object OOB writes, it is therefore essential to adjust a pointer's bounds information as soon as it starts pointing to a child object. While PCIs create a new pointer, BNIs transform an existing pointer to a pointer referencing the original object's child.

Algorithm 1 shows a high-level overview of our approach's dynamic analysis stage, and is intended to be applied to every instruction of the program upon its execution. Operations related to the core challenges solved by our design are marked in orange. For a dependent write, we identify the intended destination object from the destination pointer's taint. Using the bounds-narrowing information associated with the dependent write, we check if a write is fully in-bounds. For independent writes, the intended destination object is fixed at compile time, thus we check if the write is in bounds from the knowledge of the written bytes and the exact instruction. For PCIs and BNIs, we taint newly created pointers and re-taint existing ones to narrow their bounds according to the new pointee.

3.2   Memory Layout Extraction

Maintaining information on which objects occupy which memory regions during dynamic analysis is essential for detecting OOB writes and mapping the affected regions to their corresponding objects. To do so, we leverage DWARF debug data during static analysis. While the locations of global objects are generally specified by a fixed address, stack objects are referenced as offsets from a stack frame register. We track the program's call stack at runtime by determining the start addresses of functions and monitoring for `call` and `ret` instructions.

Compiler optimizations, which typically decrease the number of objects stored on the stack, frequently reduce the lifetime of objects in memory to one or more instruction address intervals, using the space for different purposes during

---

Algorithm 1. High-level pseudocode description of our dynamic analysis stage.

```
 1: if inst is write then
 2:     dstAddr ← getWriteDstAddr(inst)
 3:     nBytes ← getWriteBytesNum(inst)
 4:     if inst is dependent write then
 5:         taint ← getPointerTaint(dstAddr)
 6:         ido ← getObjectFromTaint(taint)
 7:         if inst is BNI then
 8:             bniTarget ← getBoundsNarrowingTarget(inst)
 9:             ido ← narrowObject(ido, bniTarget)
10:         else
11:             ido ← getObjectFromIndependentWrite(inst)
12:         if [dstAddr, nBytes - 1] is not in [ido.start, ido.end] then
13:             reportOOBw()
14: else if inst is PCI or BNI  then
15:     bniTarget ← getBoundsNarrowingTarget(inst)
16:     ptr ← getResultingPointer(inst)
17:     ido ← getPointeeObject(ptr, bniTarget)
18:     taintPointer(ptr, ido)
```

the remaining part of the function. As DWARF provides detailed information on the lifetimes of objects, we leverage this to record the location of objects not only in a spatial but also in a temporal dimension.

3.3   Intended Destination Objects Identification

For bounds-checking independent writes, we rely on the fact that their intended destination object is fixed at compile time. As high-level semantics are lost during compilation, we implement bounds-checking as a final LLVM IR analysis pass, before the translation into machine code. By leveraging the IR, we avoid directly matching independent writes to memory objects and instead take a detour as follows: (1) Identify independent writes in the IR and determine their destination variable; (2) Match each independent write in ASM to its corresponding independent write in the IR; (3) Match each destination IR variable to its corresponding object in memory.

Determining Independent IR Writes and Destination Variables. For identifying independent writes in the IR during the preliminary analysis, we focus on three typical representations of mov-family instructions in the IR: the store instruction, and the llvm.memcpy and llvm.memset intrinsics. To test whether an IR write is independent and to find the variable it modifies, we trace back its def-use chain. Starting from the write's operand that specifies the destination, we find the definition that created this pointer and repeat this procedure until there are no more unambiguous predecessors. If we end up at a local or global variable definition, we conclude the write is independent. If we encounter a BNI in the def-use chain, we keep track of the child to which the pointer is modified.

Matching Independent Writes from ASM to IR. To find the corresponding independent IR write of each independent ASM write during static analysis, we rely on line number debug information that maps instructions to the source file, line, and column at which the corresponding source code is located. While the conveyed location information is irrelevant to us, we can use its distinctive features to map write instructions in the machine code to the IR domain. In practice, however, this mapping is rarely bijective. This is caused by the inherent differences between AMD64 assembly and the IR and our disregard for certain memory-modifying instructions in ASM and the IR. As such, this constitutes a best-effort approach that occasionally fails to match an independent write.

Matching IR Variables to Memory Objects. Determining the destination object for independent IR writes is arguably the simplest step as we can match on the variable names during the preliminary analysis. If bounds-narrowing traversable definitions are encountered in the write's def-use chain, and the intended destination object is thus a child of a composite object, we leverage the information obtained while handling the BNI to determine this child.

3.4   Pointer-Creating Instructions Identification

According to our experience, two types of instructions are responsible for virtu-
ally all pointer creations in machine code generated by Clang. First, instructions
such as `lea rax, [rbp + 8*rbx - 72]` combine several arithmetic operations
and registers to compute an address. This causes the `lea` instruction to be used
for almost all cases in which a pointer relative to a stack frame boundary register
is created, as it occurs when creating a pointer to an automatic variable. Second, `mov` instructions with a source address into a static section (e.g., `mov edi,
0x409678`) are typically emitted when a pointer to a static variable is created.

3.5   Bounds-Narrowing Instructions Detection

Within compiled code, identifying the locations where bounds-narrowing is performed is challenging. For this, we again leverage the LLVM IR, where pointer
manipulation is performed with `getelementptr` (GEP) instructions. For each
of them, we test during the preliminary analysis whether the instruction transforms a composite object pointer into one of its (recursive) children. If so, we
determine the narrowed child from the instruction. We hereafter refer to this as
the *bounds-narrowing target* (BNT).

In compiled code, three types of instructions are, according to our experience,
potential candidates for BNIs that are relevant to our goal of bounds-checking
writes. First, `add` instructions, which are frequently used for offsetting pointers to
struct fields and can be bounds-narrowing if the incremented number is stored in
a general-purpose register capable of holding a pointer. Second, `lea` instructions,
which we also previously identified as pointer-creating. As the creation of a
pointer to a composite object's child requires bounds-narrowing to be performed
immediately upon pointer creation, their consideration is essential. Third, the
`mov` family, which can act as BNIs in three ways: (1) If it is a PCI to the child
of a composite static variable; (2) If a pointer to the beginning of the first child
of a composite object is created from a pointer to the composite object; (3) If it
writes to the child of a composite object, the address to which is only calculated
within the destination operand. Here, the narrowed pointer is used immediately
for writing and discarded afterwards. To match BNIs in the IR to their compiled
counterparts, we again leverage line number debug information.

3.6   Intended Pointee Objects Determination

Determining a pointer's intended pointee object (IPO) during dynamic analysis is essential for bounds-checking dependent writes. We do so by leveraging
memory layout information from debug metadata and bounds-narrowing targets
identified during the IR analysis. The need of determining an IPO arises at four
different types of instructions in our design: bounds-narrowing dependent writes,
bounds-narrowing PCIs, ordinary BNIs, and ordinary PCIs. For each type, the
required actions, depending on if the pointer is tainted, are shown in Table 2.

4   Implementation

We implemented DIVAK in 2,700 SLOC of C++ and 1,900 SLOC of Python code on top of the S2E in-vivo symbolic execution platform [7] and an LLVM pass (Fig. 3). The choice of S2E as an analysis platform was mainly motivated by its facilitation of quick prototyping in this case, minimizing the development effort for the non-core functionality of DIVAK. Furthermore, its symbolic execution capabilities provide us with the ability to perform taint analysis with effectively infinitely many taint colors by maintaining a mapping from symbolic values' internal identifiers to pointee objects. As a negative side-effect, S2E introduces significant performance overhead, which indicates that it may not be the ideal choice in real-world applications. We discuss this aspect in Sect. 6.

Preliminary Analysis. DIVAK's first stage concerns the compilation using Clang 13.0.1 and our analysis of the LLVM IR. By doing so, we get the required analysis results without modifying the program, thereby achieving non-invasiveness.

To reduce complexity, we disregard two types of BNIs: *(1) Dynamic BNIs* are those BNIs for which the (recursive) child to which a pointer is created is only determined at run time. While omitting them can cause intra-object OOB writes to remain undetected, they typically only occur when arrays of structs are indexed. *(2) Bounds-shifting instructions* serve to calculate the pointer to a sibling element in the same array, thus violating our assumption that a pointer's bounds can be narrowed by a BNI but can never be widened again. Omitting them can potentially cause legitimate dependent writes to be reported as OOB.

Table 2. Approach for determining the intended pointee object of a pointer in different scenarios. DW=Dependent Write.

	Pointer is tainted	Pointer is not tainted
BNI+ DW	Narrow pointer bounds according to target and use immediately to check write	Previously missed pointer tainting. Cannot identify IPO, hence cannot narrow bounds
BNI+ PCI	Not actually a PCI, treat like BNI	Determine pointee from BNT and memory layout, taint pointer
BNI	Narrow pointer bounds according to target and re-taint	Previously missed pointer tainting. Treat like BNI + PCI
PCI	Is a BNI but was not matched. Cannot narrow bounds	Determine pointee from memory layout, taint pointer

To alleviate poor line number debug information arising from optimizations, we attach random synthetic line numbers to independent IR writes and BNIs that are missing line numbers in the LLVM IR.

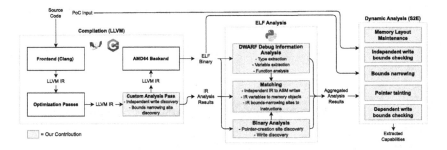

Fig. 3. Schematic diagram of DIVAK, the pipeline implementing our design.

Static Analysis. The second stage in our pipeline analyzes the compiled binary and combines the obtained results with the high-level semantic information extracted from the LLVM IR. We identify all variables and formal parameters stored on the stack or in the globals from the DWARF debug information to facilitate memory layout tracking, recording their location, lifetime, and type. We consider the smallest interval encompassing all DWARF-specified lifetime intervals as the object's lifetime. Besides, we determine the sizes of all specified types, as well as the inner structure of composite types. To address the challenges arising from representing memory lifetimes as a single interval, we merge objects with identical spatial dimensions whose lifetimes overlap.

To find pointer-creating instructions and independent ASM writes, we identify the instructions satisfying our definitions (Sect. 3). We then obtain their debug information and find their corresponding independent IR writes with identical files, lines, and column information. By finding the DWARF-specified object corresponding to each independent IR write's destination object through name-based matching, we identify the bounds of each matched independent ASM write's destination object and ease bounds-checking during dynamic analysis.

Dynamic Analysis. DIVAK runs and monitors the target program implementing Algorithm 1 in three plugins for S2E [7]. We distinguish dependent and independent writes based on their destination address. If it is symbolic, the write is dependent, and we obtain the intended destination object from the pointer's taint to test if all written bytes are within the object's interval. If it is concrete and we statically identified an independent write at this location, we check if all written bytes are part of the independent write's intended destination object.

We register callbacks that are invoked when executing instructions identified as a PCI or BNI during static analysis. When a PCI or BNI is executed, it is handled as described in Sect. 3. We determine the new intended pointee object based on the taint of the pointer, the instruction type, the bounds-narrowing information, as well as the type of the old intended pointee object. We then taint the pointer accordingly. We refrain from tainting pointers that are not stored in a register but immediately written to memory. To further improve performance, we intercept function calls to several standard library functions such as `memcpy` and perform premature bounds-checking if the destination pointer is

Table 3. Detection performance of DIVAK for dependent OOB writes in RIPE testbed at different optimization levels.

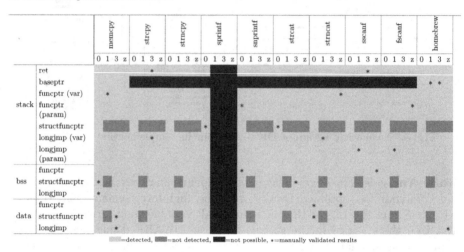

=detected, ▨=not detected, ■=not possible, *=manually validated results

tainted. Afterward, we untaint all function arguments to allow for faster emulated execution of the library function. Finally, we group OOB writes occurring at the same instruction and identical call stack, merge the overwritten intervals, identify affected source code-level objects, and report our results in a JSON format.

5   Evaluation

We evaluate DIVAK's performance in terms of OOB write detection efficacy both under laboratory conditions and in real-world programs, as well as performance overhead, and we compare it with ASan and SoftBound.

5.1   Dependent OOB Writes Detection

To measure DIVAK's ability to detect dependent OOB writes, we use a subset of a 64-bit version [28] of the RIPE testbed [39]—designed to test defenses against buffer overflow exploits. By varying the location, the type of overwritten pointer, and the function causing the OOB write, we obtain 122 test cases. By the design of RIPE, some parameter combinations are not possible. We conduct a manual best-effort validation of the affected memory objects identified by DIVAK by comparing our results with the DWARF memory layout, the memory layout to be expected from the source code, and by inspecting the disassembled code.

Results. DIVAK achieves a detection rate of 89%, as shown in Table 3. Failing test cases are limited to intra-object OOB writes under optimizations, with incomplete line number debug information for the corresponding BNI as the root

Table 4. Detection performance of ASan for dependent OOB writes in RIPE testbed at several optimization levels.

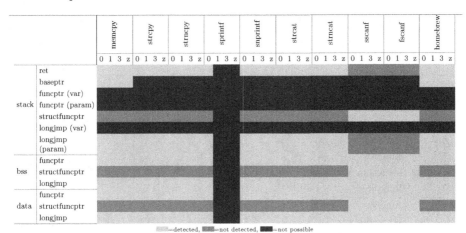

cause. Manually validating DIVAK's output yields flawless results for OOB writes in the global sections. On the stack, overwritten ranges are correctly identified. However, with optimizations enabled, the high-level counterparts of roughly 30% of the affected stack objects are not identified due to incomplete DWARF data.

ASan achieves a detection rate of 70%, as shown in Table 4. From our results, it is clear that ASan's primary drawback lies in the inability to detect intra-object OOB writes. While they were detected for `sscanf` and `fscanf`, manual analysis suggests this is caused by a bug in the testbed. SoftBound detects 34% of OOB writes. Similarly, SoftBound is limited by the inability to detect intra-object OOB writes and its reliance on six unimplemented wrapper functions.

5.2   Independent OOB Writes Detection

We evaluate DIVAK's independent OOB write detection performance using a testbed we designed for this purpose and release along with our code. The testbed comprises four parameter dimensions with a total of 44 test cases and largely reproduces the vulnerability scenarios of RIPE using independent writes.

Results. DIVAK successfully detects the OOB write in 95% of the test cases (Table 5). In 13% of the configurations, however, a false positive is detected. Detection succeeds for all inter-object OOB writes and only fails for some intra-object OOB writes on the stack and in the .`data` section. In these cases, the fault occurs in the IR analysis, where tracing the write's declaration chain terminates prematurely. Specifically, the source pointer of a bounds-narrowing `getelementptr` instruction is cast from the original structure, consisting of an array of 255 bytes and a pointer, to an array of 256 bytes. This causes the analysis to conclude that this is not a relevant BNI, as the subject is not an object we consider composite.

Table 5. Detection of independent OOB writes in our testbed. OOB writes occur in isolation (iso) or in a function containing further unrelated writes (svd).

			DIVAK								ASan								SoftBound							
			iso				svd				iso				svd				iso				svd			
			0	1	3	z	0	1	3	z	0	1	3	z	0	1	3	z	0	1	3	z	0	1	3	z
continuous	stack	ret	*				*																			
		baseptr						*																		
		funcptr																								
		structfuncptr	▪				*																			
		longjmp							*																	
	data	funcptr		*				*																		
		structfuncptr	▪		▪		▪			*																
		longjmp	*																								
	bss	funcptr																									
		structfuncptr																									
		longjmp							*																	
jumping	stack	ret																								
		baseptr	*																								
		funcptr	*				*																			
		structfuncptr		*																							
		longjmp	*																								
	data	funcptr	*																								
		structfuncptr																									
		longjmp			*																						
	bss	funcptr	*				*																				
		structfuncptr	*				*																				
		longjmp	*																								

=detected, ▪=not detected + fp, =detected + fp, ▪=not detected, *=manually validated results

Investigating the series of false positives raised by DIVAK reveals a violation of our assumption that pointers point to their intended destination object upon creation. When compiled with -O1, a pointer to a stack object is created using a `lea` but only offset to its intended pointee object by an `add` that immediately follows. Thus, the pointer is incorrectly tainted. As the integration of the `add` into the `lea` would have required fewer bytes and would presumably execute faster, the compiler's reason for splitting them remains unclear.

Manually validating the overview of affected objects generated by our approach yields similar results as the previous experiment. While the results appear correct and complete for -O0, several stack objects are missing from the summary when optimizations are employed. A closer investigation again reveals incomplete debug information generated by the compiler to be at fault.

ASan detects the OOB write in 36% of all tests. Most failures can be attributed to the inability to detect intra-object OOB writes, as well as the reliance on red zones, preventing it from detecting jumping OOB writes. Soft-

Bound successfully detects the OOB write in 73% of the test cases. All inter-object OOB writes are found, while intra-object OOB writes remain undetected.

5.3 Testing Real-World Programs

We evaluate DIVAK on three real-world vulnerabilities found in open source software. Besides, we run each program under test with a benign input to assess the false positives of DIVAK. To evaluate the performance of our static analysis, we assess the independent write and BNI matching performance. For the dynamic analysis, we collect statistics of three categories: (1) The number of dependent, independent, and unchecked writes; (2) Statistics about successful, ignored, and failed BNIs; (3) The successful pointee inferences from memory.

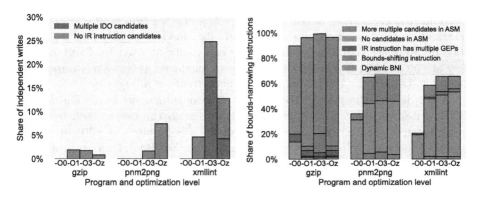

Fig. 4. Causes for unmatched independent writes.

Fig. 5. Causes for unmatched BNIs.

Table 6. Detection performance for three real-world vulnerabilities.

	libxml (CVE-2017-9047)				libpng (CVE-2018-14550)				gzip (CVE-2001-1228)			
	0	1	3	z	0	1	3	z	0	1	3	z
DIVAK												
ASan												
SoftBound												

=detected, ■=not detected + fp, ■=partly detected, ■=not possible

libxml (CVE-2017-9047). This vulnerability in the *libxml* library is a stack-based buffer overflow [4]. While attributed to the same root cause, OOB writes occur at two different instructions. As shown in Table 6, DIVAK successfully detects both OOB writes at all four optimization levels and does not yield any false positives for the PoC or the benign input. Furthermore, due to its non-invasiveness, DIVAK is the only tool to detect all occurring OOB writes. As presented in Fig. 4, independent write matching achieves a 0% failure rate without optimizations, which increases to 25% at -O3 due to overlapping memory objects and missing line number debug information. BNI matching (Fig. 5) also performs best without optimizations but has a higher failure rate, mostly due to a lack of line number debug information.

For this vulnerability, most checked writes are dependent. The share of checked writes ranges from 100% down to 31%, the latter being caused by one or more independent writes remaining unmatched. Most encountered BNIs concern pointers to unmonitored sections, mainly the heap. Last, determining a pointer's intended pointee object ranges from a 100% success rate at -O0 to a 95% success rate at -O3 for the PoC input. Most failures come from incomplete memory layout information caused by optimizations. Our assumption of memory object's lifetimes being representable by a single interval is responsible for at most 1% of failed pointee inferences. Manual validation of the affected objects identified by DIVAK yields that the results are largely correct and complete, except for a few objects that are not recorded in DWARF and thus not identified as affected.

While the OOB write affects many objects and stack frames at -O0 and -O1, rearrangement of stack objects at -O3 and -Oz causes it to only affect a single 5000-byte buffer. While the vulnerability can easily be used to divert control flow at -O0 and -O1 if no countermeasures are deployed, this is likely impossible at -O3 and -Oz if the PoC cannot be modified to affect a larger range.

ASan detects the first OOB write at each optimization level. The second OOB write, however, is not detected, likely due to the overwritten byte not being located in a red zone. Instrumenting with SoftBound at -O0 causes the compiler to crash. With employed optimizations, SoftBound reports an OOB write early during execution and crashes with a segmentation fault.

libpng (CVE-2018-14550). This stack-based buffer overflow is located in the *pnm2png* tool, part of the `libpng` library [21]. While attributed to the same root cause, OOB writes occur at two different instructions. As shown in Table 6, DIVAK successfully detects both OOB writes in pnm2png without raising false positives at any tested optimization level. For the benign input, however, false positives occur when optimizations are enabled. Manual validation of the affected objects identified by DIVAK reveals that the results are largely correct at all optimization levels, missing only an 8-byte object not present in DWARF when employing optimizations. Running an optimized version of pnm2png with a benign input using DIVAK causes multiple false positives in one function to be reported. Manually investigating the reason for this shows that the compiler optimized the zero-initialization of a struct by using a pointer to one of its fields for zeroing both the field and its sibling fields. ASan and SoftBound detect OOB writes at both locations for all tested optimization levels.

gzip (CVE-2001-1228). This is a `.bss` buffer overflow in the gzip utility. As shown in Table 6, DIVAK successfully detects the OOB write at all four optimization levels and yields no false positives for the PoC input. For the benign input, multiple false positives are reported. While independent write matching performs reasonably well (Fig. 4), BNI matching works poorly (sps 5). Only 10% are matched at -O0 and less than 1% are matched at -O3. This is primarily caused by dynamic BNIs, which comprise between 70% and 85% of all BNIs.

Furthermore, a non-negligible number of bounds-shifting instructions is found at -O3. Manual validation of DIVAK's output shows that all affected objects are correctly identified at all optimization levels. This demonstrates that the issues of an incomplete memory layout extraction are limited to the stack.

When using a benign input, three false positives are reported at -O0 and -O1, and one is reported at -O3 and -Oz. All of these are caused by our disregard for bounds-shifting instructions. In each case, a struct pointer passed to a function is used to access adjacent structs in the same array, causing DIVAK to incorrectly report OOB writes. Both ASan and SoftBound detect the vulnerability.

5.4   Performance Overhead

We evaluate DIVAK's overhead on the three real-world programs when feeding them benign inputs. We execute the programs using ASan, SoftBound, DIVAK, and natively under different optimization levels. We run each configuration ten times and consider the means. All measurements are performed on an Intel Xeon E3-1231v3 running Ubuntu 20.04. Table 7 shows the mean runtime overheads. Since our measurements indicate no substantial differences between optimization levels, we only present the results for -O1. As expected, DIVAK currently incurs a massive performance overhead (8,000 – 44,000×). ASan and SoftBound, on the other hand, incur at most a sixfold overhead. While DIVAK's performance overhead might seem to make it unusable in practice, it is important to note that little regard was given to performance during the implementation of this prototype, resulting in design decisions with a highly detrimental impact on performance. The most severe decision is the usage of S2E, which introduces a considerable overhead by executing most of the program's code in symbolic mode. We discuss a possible approach for reducing the overhead in Sect. 6.

Table 7. Performance overhead for each program at -O1.

Program	DIVAK	ASan	SoftBound
xmllint	8113×	5.4×	-
pnm2png	8592×	4.4×	3.9×
gzip	44097×	1.8×	1.5×

6   Discussion

Our experiments show that DIVAK is well capable of characterizing OOB writes. While ASan and SoftBound cannot detect intra-object OOB writes, DIVAK's logic for detecting them is not perfect. Besides the false positives raised for the real-world vulnerabilities, intra-object OOB writes are the only test cases in our testbed experiments for which detection fails. However, the former can be reduced by introducing a small number of heuristics, for example disregarding OOB writes relative to pointers created in the current function scope and not modified by pointer arithmetic. Despite these limitations, our experiments show that DIVAK's capabilities outperform ASan and SoftBound, with a false positive in the dependent write testbed being the only apparent downside.

DIVAK's main drawback is the excessive performance overhead. A promising approach to combat this is to replace the full-system emulation of S2E with dynamic binary instrumentation, e.g., Intel Pin. This would allow implementing pointer tracking by performing taint analysis through libdft [17], eliminating the overhead introduced by the use of symbolic variables. Although Pin would modify the program's memory layout by allocating space for its own metadata, the arrangement of objects within the program's sections would remain untouched. Thus, our goal of non-invasiveness would in practical terms be achieved. With the authors of libdft having measured an overhead of at most $6\times$ for typical programs, our tool would experience a substantial decrease in overhead, even with a very conservative estimate of additional $50\times$ overhead due to the amount of tainted data and our analysis logic. While DIVAK's need for a large number of taint colors would increase the load on libdft, limiting the set of colors by reusing them at the cost of a low chance for false negatives would be conceivable.

Thus, we conclude that, for use cases where only a low number of executions are necessary, e.g., bug triaging, a high recall is desirable, and false positives are tolerable, DIVAK is superior to instrumentation-based tools like ASan and SoftBound. This is especially true if intra-object OOB writes are to be detected. In addition to these benefits, it is important to keep in mind that our work's primary goal was to design a *non-invasive* OOB write detection approach that can be used to faithfully characterize the *real* effects of vulnerabilities as they exist in programs deployed in production environments: DIVAK is the only approach that guarantees faithful results in terms of affected memory objects.

Limitations. DIVAK assumes that, once the bounds attached to a pointer are narrowed to a composite child, any derived pointer requires identical or narrower bounds. This does not hold for bounds-shifting instructions, causing false positives. Nevertheless, this can be mitigated by handling such instructions akin to BNIs. A drawback of DIVAK's reliance on DWARF debug information is a dependence on its correctness and completeness. While we did not encounter cases of incorrect information, we observed incomplete location descriptions under optimizations caused by compiler bugs. This occasionally causes DIVAK to fail tainting pointers or deliver incomplete memory layout results. However, as the analysis of such bugs gained traction in the past years [2,10], enabling them to be fixed, their impact on DIVAK's results can be expected to decrease.

Another current limitation is the assumption that a single interval can describe the lifetime of any object, as outlined in Sect. 4. This occasionally causes overlapping objects, which we try to combat by merging identically-sized objects. Nevertheless, this often leaves some overlapping objects in heavily inlined code, for which we observed up to 3% of objects to overlap with one another.

Lastly, we assume the program under test to be built without frame pointer omission and tail call optimizations, arguably violating DIVAK's non-invasiveness. Furthermore, we currently do not support position-independent executables. Both issues, however, are merely limitations of our current prototype that do not invalidate our results and can be alleviated with limited implementation effort.

Future Work. A subject for future work is the extension of our design with omitted features, e.g., dynamic BNIs and bounds-shifting instructions. Moreover, re-implementing DIVAK to decrease its performance overhead is desirable to make it scalable as a part of other pipelines. Although DIVAK is meant to triage identified vulnerabilities and the discovery of new vulnerabilities is out of scope, a future research direction is the combination of DIVAK with approaches for finding alternative vulnerable paths, e.g., directed fuzzing, to create a more complete profile of the vulnerability capabilities. As the issue of invasiveness predominantly concerns the stack and globals, we do not consider heap-based OOB writes. However, DIVAK can be extended to intercept calls to memory allocators. Alternatively, one may use ASan's heap-based OOB analysis with disabled instrumentation of stack and global sections to largely maintain non-invasiveness.

7 Related Work

Several approaches have been proposed for detecting spatial memory bugs. Most of them rely on compile-time instrumentation (CTI) to insert their checking logic into the program [8,20,23,29], allowing for low overhead at the cost of highly invasive program modifications. Binary instrumentation-based approaches suffer from the lack of high-level semantic information, preventing them from providing strong spatial guarantees for the detection of certain OOB write types [31,33,37]. Similarly, binary-level pointer analysis [18] is often course-grained and cannot guarantee sufficient precision to track OOB writes that have marginal effects.

Identity-based approaches check whether the accessed memory locations are part of the expected object according to high-level program semantics. For accesses relative to pointers, this requires a sophisticated approach to maintain a mapping between pointers and intended pointee objects. *Pointer-based* approaches like DIVAK augment pointers with additional metadata, by embedding bounds information into pointers at pointer-creation sites [8,20,23,26,30]. *Object-based* approaches associate metadata only with memory objects, not with pointers [41], and test whether pointer arithmetic instructions have the same pointee before and after the operation. However, such approaches generally cannot detect intra-object OOB writes as composite objects overlap with their children.

Tripwire-based approaches [29,31] insert *red zones* around objects to detect OOB accesses. The main example of such approaches is ASan [29]. Its low-performance overhead makes it ideal for use cases such as fuzzing. A hardware-assisted variation for ARM64 [8] further decreases the memory overhead, while a kernel variation, KASAN [36] facilitates kernel fuzzing. One shortcoming is that non-contiguous OOB writes jumping over the red zone remain undetected. Furthermore, the insertion of new memory objects makes them invasive by design.

SoftBound [23] is a pointer-based sanitizer using CTI that provides relatively strong spatial detection guarantees but is heavily invasive and relies on wrappers for external function calls. While it promises to be able to detect intra-object OOB writes, we discovered this is not implemented in the publicly available tool.

Memcheck [31] and SGcheck [37] are tools for the Valgrind platform [24] and use a tripwire and heuristic-based approach, leveraging dynamic binary instrumentation without high-level semantic information. Orthogonally to our approach, QASan [13] detects heap memory violations. Intel MPX [26] is an AMD64 ISA extension that leverages CTI to get strong spatial guarantees. However, its deprecation caused its support to be removed from most compilers.

Besides in sanitizers, OOB write detection has application scenarios in larger pipelines. KOOBE [6] leverages KASAN and a pointer-based approach, similar to DIVAK, for heap-based kernel exploitation. BORG [25] discovers buffer overreads by employing a heuristic approach for recovering memory layout. Revery [38] employs a software-based memory tagging approach for heap-based AEG.

Finally, other related approaches identify memory errors at the LLVM IR level [34,35], however, they focus on different classes of bugs, such as memory leaks and use-after-free, which intrinsically require a less intrusive analysis.

8   Conclusion

We proposed DIVAK, a tool to detect OOB writes in a non-invasive manner and to distill their capabilities by identifying the affected source code-level objects stored in memory. Using two benchmarks and three real-world vulnerabilities, we showed that DIVAK can keep up with, and in some cases even exceed, the detection performance of current instrumentation-based OOB write detection approaches, yielding negligible false positives, at the cost of higher overhead.

References

1. Anderson, J.P.: Computer security technology planning study. Tech. rep., U.S. Air Force Electronic Systems Division (1972)
2. Assaiante, C., D'Elia, D.C., Di Luna, G.A., Querzoni, L.: Where did my variable go? Poking Holes in incomplete debug information. In: Proceedings of the ACM International Conference on Architectural Support for Programming Languages and Operating Systems (ASPLOS) (2023)
3. Avgerinos, T., Cha, S.K., Hao, B.L.T., Brumley, D.: AEG: automatic exploit generation. In: Proceedings of the Network and Distributed System Security Symposium (NDSS) (2011)
4. Böhme, M.: oss-security - Invalid writes and reads in libxml2 (2017)
5. Cha, S.K., Avgerinos, T., Rebert, A., Brumley, D.: Unleashing Mayhem on Binary Code. In: Proceedings of the IEEE Symposium on Security and Privacy (S&P) (2012)

6. Chen, W., Zou, X., Li, G., Qian, Z.: KOOBE: towards facilitating exploit generation of kernel out-of-bounds write vulnerabilities. In: Proceedings of the USENIX
Security Symposium (2020)
7. Chipounov, V., Kuznetsov, V., Candea, G.: S2E: a platform for in-vivo multi-path
analysis of software systems. In: Proceedings of the International Conference on
Architectural Support for Programming Languages and Operating Systems (ASP-
LOS) (2011)
8. Clang: hardware-assisted addresssanitizer design documentation (2022)
9. Cowan, C., et al.: StackGuard: automatic adaptive detection and prevention of
buffer-overflow attacks. In: Proceedings of the USENIX Security Symposium (1998)
10. Di Luna, G.A., Italiano, D., Massarelli, L., Österlund, S., Giuffrida, C., Querzoni,
L.: Who's debugging the debuggers? Exposing debug information bugs in optimized
binaries. In: Proceedings of the ACM International Conference on Architectural
Support for Programming Languages and Operating Systems (ASPLOS) (2021)
11. Ding, Z.Y., Goues, C.L.: An Empirical Study of OSS-Fuzz Bugs (2021)
12. Donovan, A.A., Kernighan, B.W.: The go programming language. Addison-Wesley
Professional (2015)
13. Fioraldi, A., D'Elia, D.C., Querzoni, L.: Fuzzing binaries for memory safety errors
with qasan. In: Proceedings of the IEEE Secure Development Conference (2020)
14. Heelan, S.: Automatic generation of control flow hijacking exploits for software
vulnerabilities, Master's thesis, University of Oxford (2009)
15. Huang, S.K., Huang, M.H., Huang, P.Y., Lai, C.W., Lu, H.L., Leong, W.M.:
CRAX: software crash analysis for automatic exploit generation by modeling
attacks as symbolic continuations. In: Proceedings of the IEEE International Conference on Software Security and Reliability (SERE) (2012)
16. ISO Central Secretary: Programming languages - C. Standard ISO/IEC 9899:2011.
International Organization for Standardization, Geneva, CH (2011)
17. Kemerlis, V.P., Portokalidis, G., Jee, K., Keromytis, A.D.: libdft: practical dynamic
data flow tracking for commodity systems. In: Proceedings of the 8th ACM Conference on Virtual Execution Environments (2012)
18. Kim, S.H., Zeng, D., Sun, C., Tan, G.: Binpointer: towards precise, sound, and
scalable binary-level pointer analysis. In: Proceedings of the ACM International
Conference on Compiler Construction (2022)
19. Klabnik, S., Nichols, C.: The rust programming language. No Starch Press (2018)
20. Kroes, T., Koning, K., van der Kouwe, E., Bos, H., Giuffrida, C.: Delta pointers:
buffer overflow checks without the checks. In: Proceedings of the EuroSys Conference (2018)
21. Luo, Z.: Stack-buffer-overflow in pnm2png in function get_token (2018)
22. MITRE Corporation: CWE Top 25 Most Dangerous Software Weaknesses (2021)
23. Nagarakatte, S., Zhao, J., Martin, M.M., Zdancewic, S.: SoftBound: highly compatible and complete spatial memory safety for C. In: Proceedings of the ACM
Conference on Programming Language Design and Implementation (PLDI) (2009)
24. Nethercote, N., Seward, J.: Valgrind: a framework for heavyweight dynamic binary
instrumentation. In: Proceedings of the ACM Conference on Programming Language Design and Implementation (PLDI) (2007)
25. Neugschwandtner, M., Comparetti, P.M., Haller, I., Bos, H.: The BORG:
nanoprobing binaries for buffer overreads. In: Proceedings of the ACM Conference on Data and Application Security and Privacy (CODASPY) (2015)
26. Oleksenko, O., Kuvaiskii, D., Bhatotia, P., Felber, P., Fetzer, C.: Intel MPX
explained: an empirical study of intel MPX and software-based bounds checking
approaches (2017)

27. PaX Team: Address Space Layout Randomization (2001)
28. Rosier, H.: ripe64 (2019). https://github.com/hrosier/ripe64
29. Serebryany, K., Bruening, D., Potapenko, A., Vyukov, D.: AddressSanitizer: a fast address sanity checker. In: Proceedings of the USENIX Annual Technical Conference (2012)
30. Serebryany, K., Stepanov, E., Shlyapnikov, A., Tsyrklevich, V., Vyukov, D.: Memory tagging and how it improves C/C++ memory safety (2018)
31. Seward, J., Nethercote, N.: Using Valgrind to detect undefined value errors with bit-precision. In: Proceedings of the USENIX Annual Technical Conference (2005)
32. Shoshitaishvili, Y., et al.: Rise of the HaCRS: augmenting autonomous cyber reasoning systems with human assistance. In: Proceedings of the ACM SIGSAC Conference on Computer and Communications Security (CCS) (2017)
33. Slowinska, A., Stancescu, T., Bos, H.: Body armor for binaries: preventing buffer overflows without recompilation. In: Proceedings of the USENIX Annual Technical Conference (2012)
34. Sui, Y., Xue, J.: SVF: interprocedural static value-flow analysis in LLVM. In: Proceedings of the ACM International Conference on Compiler Construction (2016)
35. Sui, Y., Ye, D., Xue, J.: Static memory leak detection using full-sparse value-flow analysis. In: Proceedings of the International Symposium on Software Testing and Analysis (2012)
36. The kernel development community: The Kernel Address Sanitizer (KASAN) - The Linux Kernel documentation (2021)
37. Valgrind Developers: SGCheck: an experimental stack and global array overrun detector (2012). http://valgrind.org/docs/manual/sg-manual.html
38. Wang, Y., et al.: Revery: from proof-of-concept to exploitable. In: Proceedings of the ACM SIGSAC Conference on Computer and Communications Security (CCS) (2018)
39. Wilander, J., Nikiforakis, N., Younan, Y., Kamkar, M., Joosen, W.: RIPE: runtime intrusion prevention evaluator. In: Proceedings of the Annual Computer Security Applications Conference (ACSAC) (2011)
40. Xu, L., Jia, W., Dong, W., Li, Y.: Automatic exploit generation for buffer overflow vulnerabilities. In: Proceedings of the IEEE International Conference on Software Quality, Reliability and Security Companion (QRS) (2018)
41. Younan, Y., Philippaerts, P., Cavallaro, L., Sekar, R., Piessens, F., Joosen, W.: PAriCheck: an efficient pointer arithmetic checker for C programs. In: Proceedings of the ACM Symposium on Information, Computer and Communications Security, (ASIACCS) (2010)

Flow Integrity and Security

CEFI: Command Execution Flow Integrity for Embedded Devices

Anni Peng[1,3], Dongliang Fang[2,3], Wei Zhou[3,4], Erik van der Kouwe[3,5],
Yin Li[1,3], and Yuqing Zhang[1,3,6(✉)]

[1] National Computer Network Intrusion Protection Center, UCAS, Beijing, China
zhangyq@nipc.org.cn
[2] Institute of Information Engineering, CAS, Beijing, China
[3] School of Cyber Security, UCAS, Beijing, China
[4] School of Cyber Science and Engineering, HUST, Wuhan, China
[5] Vrije Universiteit Amsterdam, Amsterdam, The Netherlands
[6] School of Cyberspace Security, Hainan unversity, Haikou, China

Abstract. As embedded devices are widely used in increasingly complex settings (e.g., smart homes and industrial control systems), one device is usually connected with multiple entities, such as mobile apps and the cloud. Recent research has shown that *privilege separation* vulnerabilities, which allow violations of authority between different entities, are occuring in IoT systems. Because such vulnerabilities can be exploited without violating static control flow and data flow, existing CFI and DFI solutions cannot prevent them. We present *CEFI*, the first method to enforce integrity of command execution on embedded devices after deployment. *CEFI* provides fine-grained *Command Execution Flow Integrity* by preventing external commands from being executed on control flow paths belonging to interaction channels that are not authorized to perform them. Using minimal manual annotations as a starting point, *CEFI* statically determined the legal path set (from the start to the end point) and instruments the program to verify the legitimacy of the command execution at runtime by checking whether the calling context is consistent between the runtime executed path and statically obtained legal path set. We evaluate our prototype with five real-world firmware samples, and show that *CEFI* has an average performance overhead of just 0.18%, an average memory overhead of 0.19%, and that *CEFI* can effectively protect embedded devices against attacks on privilege separation vulnerabilities even if they do not violate control flow.

Keywords: Internet of Things (IoT) · embedded devices · enforcement

1 Introduction

With the development of the Internet of Things (IoT), the application scenarios of embedded devices are becoming broader and more complicated. For example, embedded devices have long been restricted to closed environments, such as

D. Gruss et al. (Eds.): DIMVA 2023, LNCS 13959, pp. 235–255, 2023.
https://doi.org/10.1007/978-3-031-35504-2_12

industrial plants and vehicle communication systems, but nowadays are increasingly connected, communicating with external systems to carry out their own functionality. Embedded systems often communicate with multiple external systems in different roles. For example, a smart watch might communicate with one server over WiFi to receive updates, with another over a 4G cellular network to share location data, and also handle diagnostics and configuration commands received by traditional SMS. The attack surface has greatly increased over time. Attacks can send malicious data to an interaction channel by exploiting many low-level security bugs (such as buffer overflows like CVE-2020-25066, CVE-2020-27337, CVE-2020-27338, etc.) in firmware. Furthermore, attackers also leverage missing checks among different interaction channels to perform unauthorized functions.

Taking a smart home scenario as an example, a smart lock interacts with the IoT cloud and mobile app simultaneously. Specifically, the smart lock can receive operation commands both from the remote cloud and the local mobile app. The remote cloud can control the smart lock to update its firmware, and the local app can control the smart lock to perform lock operations (e.g., lock or unlock the door). Typically, different interaction channels are designed to serve different purposes. For this example, firmware updates can *only* be initiated from the trusted cloud. However, if firmware fails to properly verify its interaction channels, a local attacker can issue a malicious firmware update command to the device. This type of attack has been demonstrated in the previous research, which has uncovered 69 similar bugs [30]. Even worse, such an attack is stealthy, as a firmware update is considered to be a normal device operation. The received command does not deviate from normal ones, and there is no violation from the viewpoint of control-flow integrity. Although there is a large body of research on protecting low-end embedded devices [1,6,7,16,20,24], they only focus on basic security properties like control-flow integrity and data-flow integrity. Most recently, OAT [24] provides an attestation method that prevents both control-flow and data-only attacks on embedded devices. However, newly discovered hazards [30,35,36] involved in IoT interaction channels have enlarged attack surfaces of embedded devices. Moreover, as seen in the example, such attacks do not carry abnormal data or violate the control flow, so none of the previous works can detect such attacks on the device side. Note that this attack differs from data-only attacks [13], which usually require the exploitation of memory corruption bugs. The root cause of our example is a *logical flaw* in the design or implementation of the product, which remains unknown to both vendors and users. It is different from the issue of implicit authorization, as discussed in SmartAuth [26]. Implicit authorization refers to the mobile app gaining more privileges without notifying the user, which is a problem that vendors are aware of but users are not. This also differs from research that primarily analyzes the mobile applications to identify and exploit potential security issues, as seen in studies such as [11,12]. Although there has been considerable research on security issues in mobile applications, we found mitigating logic flaws in IoT embedded devices have not yet been systematically studied in the literature.

In this work, we propose the first interaction command based attestation method that verifies whether the requested operation is trustworthy when it carries out one received command execution, even when strictly following its designed purpose. We automatically enforce this verification even if part of the call stack is shared between command handlers, requiring only minimal manual annotations to indicate which commands are allowed on which interfaces. We assign a unique code (i.e., an integer value) to each different code path, and automatically generate and enforce an allowlist that specifies all legal code paths. To prevent attackers from manipulating the unique code and the allowlist, we store both in secure memory on ARM-based devices using on the widely deployed TrustZone extension. This allows us to prevent attackers from executing commands from contexts where they are not authorized, even if executing them would not violate control-flow and data-flow integrity.

Contributions. Our work makes the following contributions:

- We propose <u>C</u>ommand <u>E</u>xecution <u>F</u>low <u>I</u>ntegrity (CEFI), the first method to enforce integrity of command execution on embedded devices after deployment, even against attacks that violate neither static control flow nor static data flow.
- We apply a calling context encoding algorithm to classify each unique control flow of the program, which is lightweight and suitable for resource-constrained embedded devices.
- We implement *CEFI* and conduct evaluation over five real-world embedded programs that broadly cover multiple use cases in IoT devices, demonstrating the practicality of *CEFI* in real-world application scenarios. *CEFI* is available at https://github.com/mituanzi/CEFI.

2 Background

2.1 IoT Architecture

IoT architectures typically involve multiple types of entities [35,36] including IoT device, cloud backend, and the companion mobile apps running on smartphones (see Fig. 1). Each entity has different responsibilities and design goals. The *IoT device* is designed to interact with the physical world through sensors and actuators. It sends collected real-time information (e.g., device status and environment events) to the cloud or the mobile application. The *cloud backend* manages devices and mobile app user accounts, including the binding between devices and user accounts. In addition, device firmware can be updated from the cloud when needed. When users are not in the same LAN with the devices, the cloud can act as a proxy to forward device control commands from remote users, and forward the device status or command execution results back to the app. *Mobile apps* provide users with an interface to manage devices (e.g., binding the device, viewing device status, and issuing control commands). Generally, mobile apps can control the device in two ways: 1) directly send the commands to the device if they are in the same LAN, or 2) indirectly send the requests to

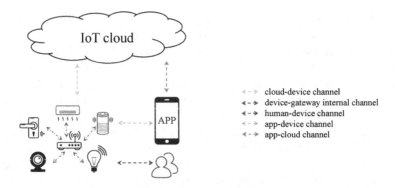

Fig. 1. Interaction model of IoT platform

the device via the cloud remotely. There are bidirectional interaction channels between each pair of entities (see Fig. 1).

2.2   Interaction Channels on IoT Platform

In the IoT platform, each entity plays a different role and takes on different responsibilities. As a result, the interaction channels between these entities present a certain complexity. Specifically, i) There are many interaction channels derived from multiple entities. First, the relationship between the entities is not only in a one-to-one pattern, but also a one-to-many or many-to-many pattern. For example, a device can be accessed by multiple different users and delegated to multiple different third-party platforms [31] (e.g., Philips Hue, LIFX, Google cloud, etc.). Second, entities may also have many interactions inside. For example, a smart hub can interact with multiple smart lights via ZigBee. Finally, a human may also participate in the interaction model directly and bring new interaction channels, such as controlling the device based on the human voice or directly controlling the device with a physical touch screen. ii) Each channel has a different design purpose, which is mentioned in Sect. 2.1. In order to suit the situation, the interaction channels involve a variety of different communication protocols, such as Bluetooth, Zigbee, MQTT, HTTP, etc. iii) There may be functional overlap between different interaction channels. For example, both the cloud and the mobile app have the same functionality in controlling the device (e.g., turn on/off the device), recall that the cloud can act as a proxy to forward control commands from the mobile app. We also note that the app→device channel and cloud→device channel have different responsibilities to manage the device account [30].

The complex interaction model among multiple entities in IoT platforms makes maintaining design purpose more complicated, increasing the security risk of many critical tasks, such as authentication [22,36], privilege separation [15,30, 31], state synchronization [21], and task isolation [14], etc. This paper focuses on mitigating *Privilege Separation Vulnerabilities* (PSVs) that violate the privilege separation model [30] (see Sect. 3.1).

2.3 ARM TrustZone

Our system relies on ARM TrustZone to protect critical state. TrustZone offers a trusted execution environment on ARM. It is available even on Cortex-M microcontrollers, which allows our system to be used even on low-cost CPUs optimized for ultra-low power embedded applications.

TrustZone introduces two protection domains with different permissions at the processor level, the Secure World and the Normal World. The two worlds are completely isolated by the hardware and have different permissions. While code executed from Secure World can access memory in both secure and non-secure regions, both applications and operating systems running in the Normal World are prevented from accessing the resources of the Secure World. Access is only possible through API interfaces specifically offered for this purpose by software in the Secure World. Properties such as hardware isolation and different permissions between the two worlds provide an effective mechanism for protecting an application's code and data, even in the face of a compromised operating system kernel. Therefore, TrustZone can be used to protect the state that is critical for security of our system from tampering.

3 Motivation

3.1 Problem Statement

In this section, we show the dangers of privilege separation vulnerabilities, and the need for a lightweight system to mitigate them, especially on low-powered IoT devices. Following Yao et al. [30], we define privilege separation vulnerabilities as vulnerabilities that violate the privilege separation model. The privilege separation model defines the privileges of each involved role (e.g., remote cloud, local app), which can be inferred from the specification, program context, empirical knowledge, etc.

In order to explain the vulnerabilities in detail, we abstracted a piece of pseudocode from a real example, as shown in Listing 1.1. It demonstrates two types of privilege separation vulnerabilities, based on examples in prior work [30]. The code implements two independent handlers (i.e., cloud_handler and local_handler) to process data from the remote cloud and the local app respectively (line 5). Both handlers use similar processing logic (line 10-39): they receive data over a network, perform authentication, and parse the data. Both invoke the `extract_cmd` function whenever a command is specified. However, `cloud_handler` and `local_handler` may use different receive functions (e.g., `ssl_recv` and `tcp_recv`), different protocols (e.g., MQTT and HTTP), and different data formats (e.g., encrypted format and JSON format). The `extract_cmd` function extracts the specific command and its parameters, and after some checks (e.g., format and value range checks), it executes the command by invoking the corresponding execution functions.

```
1   void task_main(void) {
2     /* initializations (e.g., variable declarations  */
3     /* and definitions, memory allocations, etc) */
4     // ......
5     func_t handlers[2] = {cloud_handler, local_handler};
6     /* register remote handler and local handler */
7     // ......
8   }
9
10  void cloud_handler(void) {
11    /*  initializations */
12    // ......
13    char *buf = remote_recv(); // e.g., ssl_recv
14    if (auth(buf)) {
15      parse_remote_data(buf);
16      /* check whether it contains command */
17      // ......
18      if (remote_control) {
19        extract_cmd(buf);
20      }
21      // ......
22    }
23    /* error handling and response */
24  }
25
26  void local_handler(void) {
27    /*  initializations */
28    char *buf = local_recv(); // e.g., tcp_recv
29    if (auth(buf)) {
30      parse_local_data(buf);
31      /* check whether it contains command */
32      // ......
33      if (local_control) {
34        extract_cmd(buf);
35      }
36      // ......
37    }
38    /* error handling and response */
39  }
40
41  void extract_cmd(char *buf) {
42    cmd_t *cmd = parse_command(buf);
43    parameter_t *para = parse_parameter(buf);
44    /* check whether cmd and parameters are valid */
45    /* (e.g., check its format, value range, etc) */
46    // ......
47    switch (cmd.type) {
48      case OP_1_TYPE: exec_turnOn(cmd, para);   break;
49      case OP_2_TYPE: exec_update(cmd, para);   break;
50      case OP_3_TYPE: exec_turnOff(cmd, para);  break;
51      case OP_4_TYPE: exec_reboot(cmd, para);   break;
52      // ......
53    }
54    // .....
55  }
```

Listing 1.1. Simplified Code Snippet on interaction channel processing logic.

By design, the IoT devices perform privileged operations that can only be initiated from specific interaction channels. Therefore, IoT devices should implement a strict *privilege separation model* when handling commands from different channels (entities) [30]. For example, commands from the remote cloud are

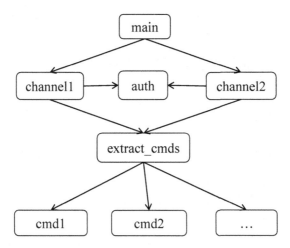

Fig. 2. Simplified Call Graph Illustration

responsible for device management, such as binding or unbinding the device with the owner, and updating the device's firmware. Commands from the local apps are used to control interaction with the environment (physical world), e.g., turning on/off the switch, locking/unlocking the lock, adjusting the brightness of the light, etc. Different channels are not supposed to interfere with other's responsibilities. For example, the local app should not be able to update the device's firmware. These properties can be violated by PSVs. We will discuss two particular types: over-privilege vulnerabilities and authentication bypass vulnerabilities.

Over-Privilege Vulnerabilities. Listing 1.1 shows an *over-privilege* vulnerability. The command handling function `extract_cmd` shared between the interfaces supports all commands, even those not authorized on some of them. There is a "valid" execution path (that is, one not violating CFI properties) from `local_handler` to `exec_update`, namely with the call trace `task_main` → `local_handler` → `local_recv` → `auth` → `parse_local_data` → `extract_cmd` → `exec_update`. However, based on the program context, it appears that the "valid" path is unexpected and violates the privilege separation model. We can infer this from the user app lacking a user interface (e.g., button) that can initiate the exec_update behavior. This behavior is not intended for the user app to perform. However, attackers can bypass the app interface and issue the command using scripts. This bug has been reported to the vendor and acknowledged [30]. Inferring expected behavior from the program context, including exposed user interfaces, is already used in existing research [26].

In this paper, we label such bugs as *over-privilege* vulnerabilities. Over-privilege vulnerabilities are common in IoT platforms, and previous research has

```
1    int auth(char *buf) {
2        /*extract user and password information from the buf*/
3        // ......
4        if ((strcmp(user, "GO") == 0) &&
5            (strcmp(pass, "ON") == 0))
6            return SUCCESS;
7        // ......
8        /* this is the real auth function, it checks */
9        /* user provided data with the credentials */
10       if (real_auth(user,pass))
11           return SUCCESS;
12       else
13           return FAIL;
14   }
```

Listing 1.2. Code Snippet from [22] with slight changes.

uncovered many severe bugs [30,35,36]. Over-privilege are at the root of a number of CVEs, including CVE-2018-10691, CVE-2020-26072, CVE-2022-36782, and CVE-2022-41627. To better understand the root cause of over-privilege vulnerabilities, we show a simplified call graph (see Fig. 2) abstracted from Listing 1.1. We assume there is a design goal that cmd1 can only be reached through channel1, and cmd2 can only be reached by channel2. From the graph, we can see that there are two unintended paths: channel1 can reach cmd2, and channel2 can reach cmd1. When taking a closer look at its root cause, we have two key observations: i) before the command execution functions (i.e., cmd1, cmd2) there is no check on the role or channel that issues the command, and ii) the unintended path uses a shared function extract_cmds to dispatch commands. This function mixes the relationship between channels and privileged operations.

Authentication Bypass Vulnerabilities. In the Listing 1.1, the auth() function can also be bypassed without having knowledge of user's credentials. Generally, attacker may leverage control-flow hijack technique (e.g., based on some memory bugs) to bypass the authentication process. However, we do not target memory bugs in this paper, instead, we target the bugs that are derived from logic violations – the presence of hardcoded authentication credentials in the authentication routine. Specifically, we show a problematic implementation of auth() function in the Listing 1.2. There are hardcoded credentials ("GO" and "ON") in the auth function. The auth function can be bypassed (i.e., without calling the real_auth function) when the user's input is consistent with the hard-coded ones (line 2-4). Consequently, once an attacker analyzes and knows the relevant content of the hardcoded information, it is not hard for them to bypass the authentication and gain access to the device. Authentication bypass vulnerabilities are common, and are at the root of a number of CVEs, including CVE-2017-8226, CVE-2021-33218, CVE-2021-33220, CVE-2022-29730.

Limitations of Potential Solutions. To defend against the privilege separation bugs, there are three potential solutions: i) Detection in advance by analyzing illegal path reachability. Ideally, we can use static analysis to detect illegal paths in advance and ensure that cmd1 will never be reached through channel2. However, since static analysis faces some common challenges (e.g., it is hard to precisely resolve all indirect calls), it is very hard if not possible to accurately exclude all paths from channel2 to cmd1. Moreover, any detected bugs still need code patches (defense solutions) to defend against potential attacks. ii) Blocking execution by checking input directly at the start point channel2. Recall the example in Listing 1.1, one may argue that we can easily block the execution if an "update" command is found at the local_recv(). We note that the received raw data at the entry point has various complex formats (e.g., JSON format and encrypted formats [30]) and not completely parsed yet at this point. Therefore, it is infeasible to directly filter illegal commands at the entry point (i.e., local_recv() and remote_recv()). Futhermore, different channels are not completely independent and often have some shared behaviors (e.g., both the cloud and user app are allowed to issue turn on/off commands) and call shared functions. Shared functions can easily lead to privilege separation vulnerabilities, as noted by Yao et al. [30]. Although the program can know which channel is involved at the entry point, it cannot predict the control flow, since the command has not yet been parsed at that point. iii) Applying traditional CFI solutions. However, logic bugs follow a "legal" path in the program implementation. There is a path from channel2 to cmd1 which does not violate the CFG, so CFI solutions fundamentally cannot mitigate this type of vulnerability.

3.2   Threat Model

We focus only on privilege separation vulnerabilities, and assume that the attacker cannot hijack the original control flow and data flow. Note that we do not consider memory errors in the threat model, as existing work can mitigate them. Our approach is orthogonal to existing solutions addressing control flow hijacking (e.g., [1,20]) and data-flow violations (e.g., [24]). We also assume that the attacker can access the IoT devices (e.g., via victim's LAN), so they have the ability to send the requests to the IoT devices directly or with a MITM attack. Moreover, we assume the attacker can exploit any logic errors to execute commands already present in the firmware without authorization. Our system will be applied by a programmer who has access to the firmware source code, and has knowledge of its high-level privilege-related design logic. This is realistic when our system is deployed by the original developer, and also for third parties in case of properly documented open source firmware. We also assume the trusted software in the TEE is bug-free and isolated from the Normal World firmware, as important metadata such as the allowlist is stored there. Considering the small code base of the TEE-side software and the limited attack surface, this is a reasonable assumption, well accepted by existing TEE-based work [1,20,24]. We do not consider low-level physical attacks, such as connecting to a JTAG debugger to re-program the firmware. Finally, we assume our compiler passes are free of bugs.

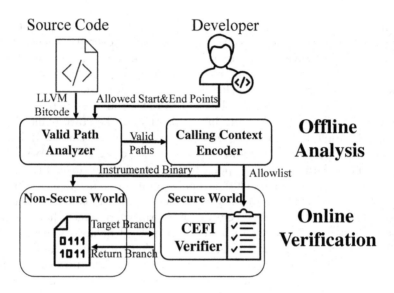

Fig. 3. The architecture of CEFI.

4   CEFI

In this section, we discuss the design and implementation of *CEFI*. Figure 3 shows the overall design of *CEFI*. Developers can use *CEFI* to protect their firmware. It acts as a compiler pass, and anyone who has access to the firmware source code can use it to generate a firmware binary hardened against privilege separation vulnerabilities. *CEFI* needs minimal manual annotations to specify the mapping between interfaces that can receive commands and commands permitted on those interfaces, and can then automatically instrument the program to enforce those policies at runtime, even in the face of logic bugs.

Our approach consists of two phases, which are discussed in this section. The static calling context encoding (CCE) phase (Sect. 4.1) happens at compile time, and performs static analysis and instrumentation to create a hardened binary. It uses the policies specified by the user in the form of minimal annotations to generate an `allowlist` that specifies the code paths that satisfy the policy, and are therefore valid contexts to execute particular commands, and instruments to code to be able to enforce this allowlist. The dynamic command execution flow verification phase (Sect. 4.2) offers the necessary support at runtime to perform these checks in a secure way. We take advantage of secure storage and isolation provided by Trustzone-M the `allowlist` at runtime. We use Trustzone to protect the security of the allowlist, as our approach relies on the guarantee that it is not illegally written to. Meanwhile, due to the security isolation of TrustZone, we avoid having to perform checks at all risky (e.g., pointer-based) write instructions, which is expensive.

4.1   Calling Context Encoding Instrumentation

We implement *CEFI* as two LLVM compiler passes, namely the Valid Path Analyzer and the Calling Context Encoder. We discuss them in this section.

Valid Path Analyzer. To accurately enforce the execution path belonging to its expected interaction channel, *CEFI* needs developers to specify a legitimate pair set between the sensitive command function (i.e., end point) with its corresponding entry function (i.e., start point). There are some manual works annotating the start point and the end point, and representing their relationship. Taking Listing 1.3 as an example, function `func_1` and function `func_2` are two different entry functions (i.e., start point). The function `turn_on` and `update_firmware` are two different command execution functions (i.e., end point). The annotations specify a relationship between the start points and the end points, indicating which paths are allowed and which are not. Specifically, the end point `turn_on` can be reached from start point `func_1` and `func_2`, while end point `update_firmware` can only be reached by start point `func_1`. On this basis, we can obtain many allowed pairs, i.e., (`func_1`, `update_firmware`), (`func_1`, `turn_on`), (`func_2`, `turn_on`). After annotating, the path analyzer gathers all available paths from the start point to the end point of each pair based on the call graph with a function named `AllowPathAnalyzer()`. In this way, we can get the valid paths to each command execution function and filter out a lot of irrelevant code. Note that these valid paths have already filtered the paths that are not allowed to reach command execution functions, e.g., any paths from `func_2` → `update_firmware` are not regarded valid even if they exist.

```
1    void __attribute__((annotate("entry#role1"))) func_1();
2    void __attribute__((annotate("entry#role2"))) func_2();
3    void __attribute__((annotate("cmd#role1#role2"))) turn_on();
4    void __attribute__((annotate("cmd#role1"))) update_firmware();
```

Listing 1.3. Manual Annotation

Calling Context Encoder. Calling context encoding is a lightweight technique to record dynamic calling path history, which has been widely used in many software development processes such as testing, event logging, and program analysis [4,23,32,33]. Its basic idea is to instrument function calling point, so at the runtime, the instrumentation can dynamically update the ID such that the value of the ID represents the current calling context. To make the ID uniquely distinguish different contexts, the CCE algorithm solves many challenges, such as recursive calls, function pointers, etc. Since the CCE algorithm is not our contribution, we omit the details here. Instead, we directly integrate the work by Sumner et al. [23] in *CEFI*. The left-hand side of Fig. 4 (see [23] for the ID notation) illustrates how it works. In the calling graph, the CCE algorithm assigns a value to each edge between `StartPA` and `EndPA`. so the ID can be dynamically updated and each path to EndPA can be uniquely mapped to a value in the runtime. Specifically, before the start point StartPA, the id is initialized to 1, and at the end point EndPA, the id may have two different values (i.e., 1, 2), which can uniquely distinguish two different paths ($StartPA{\rightarrow}f2{\rightarrow}f4 \rightarrow EndPA$ and $StartPA{\rightarrow}f3{\rightarrow}f4 \rightarrow EndPA$). When $CEFI$ is deployed, it will instrument a secure gateway API call at each edge, but it only requires instrumentation at the edge $f3{\rightarrow}f4$, as the process of "$id+ = 0$" does not alter the calling context ID, making instrumentation unnecessary at other edges (such as StartPA$\rightarrow$f2, StartPA$\rightarrow$f3, and f2$\rightarrow$f4). $CEFI$ is lightweight, which can be attributed to the fact that only a limited amount of instrumentation is required. Besides, the involved CCE algorithm is safe (i.e., different contexts are guaranteed to have different ID), reversible (i.e., calling context can be faithfully decoded and recovered) [23].

With the help of CCE, we can generate an allowlist for each end point (i.e., command execution function), as shown in Algorithm 1, which is denoted as AllowlistGenerate(). The algorithm takes the pair set of user-defined start and end points and the whole call graph of the firmware as inputs. The output is the allowlist. For each pair (i.e., StartPA, EndPA), the algorithm first gets an AllowPathSet using the ValidPathAnalyzer(), and each item in the set represents an allowed path from StartPA to EndPA. Then, it encodes each edge of an allowed path using CallingContextEncoding(), and accumulates the weights (IDs) of each allowed path to EndPA using ComputeID. Specifically, each call statement on the allowed path will be instrumented to update the value of context identifier, so that we can obtain a value at the EndPA and check it against the allowlist. Note that we assign one to the start point instead of zero as the traditional CCE algorithm does for quickly distinguishing the remaining paths to the command function with zero values without encoding them. After traversing every *item* in *pairSet*, we obtain the allowlist dictionary, in which the key is EndPA and the value is AllowedIDset.

Instrumentation with ARMv8-M Security Extension. During instrumentation, the firmware reads the allowlist and resides in the read-only secure memory before initialization at runtime. Then, all allowed paths are encoded with CCE as mentioned before. Meanwhile, any ID update before the function call site will be transferred to TrustZone-protected Secure World, thus we can guarantee the security of dynamically computed calling context ID. To enforce the integrity check before command execution, we also intercept all call instructions to the command functions (end points) and redirect them to the CEFI verifier in the Secure World through secure gateway veneers [3]. If the verification is passed, the control flow returns back to the intended command function.

Algorithm 1. Algorithm for Allowlist Generation, denote as AllowlistGenerate()

Input : PairSet
Input : CallGraph
Output: Allowlist
//PairSet: each item consists of (startPoint, endPoint)
//StartPA: entry function
//EndPA: command function
//Allowlist: a dict, the key is EndPA, the value is an AllowedIDSet
Item ← Head(PairSet)
do
    StartPA, EndPA ← GetTuple(Item)
    AllowPathSet ← ValidPathAnalyzer(StartPA, EndPA, CallGraph)
    EncodedAllowPathSet ← CallingContextEncoding(AllowPathSet)
    AllowedIDSet ← ComputeID(EncodedAllowPathSet)
    Allowlist ← AddTo(Allowlist, EndPA, AllowedIDSet)
    Item ← Next(PairSet);
while *Item*;

4.2 Command Execution Flow Integrity Enforcement

Figure 4 illustrates the process of dynamic command execution flow integrity enforcement with CEFI verifier. When a call instruction to a command function is encountered in the firmware, the control flow will be transferred to the security gateway veneers and further forwarded to the CEFI verifier. It verifies the current calling context (current ID value) by looking up the valid ID set to the command function (e.g., EndPA in Fig. 4) from the allowlist. After verification, if the calling context is allowed, the control flow redirects back to the command execution. If the calling context is invalid (e.g., the ID value is not one or two in Fig. 4), the CEFI verifier will stall the execution and output the current calling context (ID value) via UART or other peripherals based on user specification.

5 Evaluation

To evaluate *CEFI*, we conducted experiments on five commonly-used programs to i) measure its performance in terms of runtime and memory overhead, ii) perform a security analysis to demonstrate its effectiveness, and iii) show the manual effort required to use *CEFI*.

Testing Environment. *CEFI* is implemented on top of LLVM 10 and runs on Ubuntu 18.04. *CEFI* generates hardened binaries that can run on the STM32L562E-DK discovery kit, a popular IoT development board. This board features an ARM Cortex-M33 core with TrustZone support, along with 512 kB Flash memory and 256 kB SRAM. Our prototype does not rely on other board-specific features, making it adaptable to other ARM Cortex-M chips.

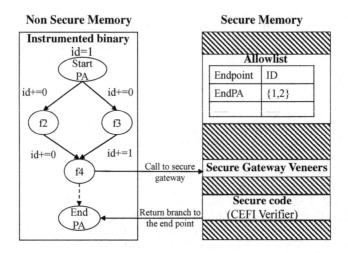

Fig. 4. A Running example of the CEFI verification process.

Benchmark. Our benchmark consists of five embedded programs with practical application scenarios. They are small in size (the average size is 104.03 kB) compared with traditional software, which is representative of embedded programs in practical settings, as these CPUs cannot run larger programs. These programs have been used for evaluation in prior embedded device research [8,10,24].

- **Light Controller** is used in smart home applications. User can turn the light on/off remotely by sending control command. It is also used for evaluation by OAT [24].
- **Syringe Pump** is used in medical and production applications. The user can control a device to inject or withdraw fluid automatically by sending control command with user-provided amount. It is also used by C-FLAT [1] and OAT [24].
- **Thermostat** reads the temperature and humidity from a sensor. If the temperature is too far from a preset temperature it can, for example, trigger an air conditioning unit. It also accepts commands to retrieve the current temperature. This program is used by PRETENDER [10].
- **RF_door_lock** can be applied to smart door locks. Its commands include unlocking the door given the correct password, and setting a custom password. This program is used by PRETENDER [10].
- **Steering_control** is used in autonomous driving. It receives commands from the computer to control the steering and moving/motoring of the autonomous vehicle. This program is used by P2IM [8].

Table 1 presents details on the application layer logic of the programs. It does not consider the boot loader, board support package (BSP), device drivers, or any other components outside of the application layer. It includes the functions involved (#Functions), the number of annotations made (#Annotations), the number of all allowed paths from the StartPA to EndPA (#AllowedPaths), and the lines of code of application layer (#LoC).

Table 1. Statistics in Benchmark Programs

Program	#Functions	#Annotations	#AllowedPaths	#LoC
Light Controller [24]	33	4	7	286
Syringe Pump [1, 24]	51	4	8	569
Thermostat [10]	28	5	4	154
RF_door_lock [10]	25	4	5	219
Steering_control [8]	33	2	8	150

5.1 Performance Overhead

For defense mechanisms to be deployable, they must result in low performance overhead [25]. This is especially important for resource-constrained embedded devices. To study the runtime overhead introduced into a system by *CEFI*, we measure the execution time with and without *CEFI* for each test program. We record end-to-end overhead, based on the time between when a device receives an event and when a device completes the resulting action. The five programs we selected all have multiple execution paths that are triggered by different inputs. Therefore, we design different inputs to trigger each branch of the program. The reported runtimes are averages over ten executions of each input.

We present the runtime overhead and memory overhead in Table 2. The column #Trans lists the average transitions between the trusted and normal world of each programs. The results show that *CEFI* has very low overhead for each of the programs, with a geometric mean of just 0.18% over all of them.

Memory overhead consists of Flash overhead and RAM overhead. For Flash overhead, *CEFI* adds instrumentation to encode and decode the ID at each call site. In addition, before the specific function, we need instrumentation to send the ID to the Secure World to match it against the allowlist. These instrumentations

Table 2. Runtime Overhead and Memory Overhead

Program	#Trans	Execution Time (ms)			Memory Consumption (bytes)		
		Baseline	*CEFI*	Overhead	Baseline	*CEFI*	Overhead
Light Controller	9	18.49	18.51	0.11%	102888	103172	0.28%
Syringe Pump	7	54.34	54.36	0.04%	110536	110772	0.21%
Thermostat	5	5.04	5.05	0.20%	108184	108332	0.14%
RF_door_lock	10	2.20	2.21	0.45%	102308	102480	0.17%
Steering_control	7	10.74	10.75	0.09%	108724	108888	0.15%
Geometric Mean				0.18%			0.19%

Table 3. Commands for Smart Light Example

Interaction Channel	Annotated Command Functions
Local client	switch_on, switch_off
IoT cloud	update_firmware, recovery_firmware, change_password

increase Flash usage. The results show the Flash memory overhead of *CEFI*. The geometric mean overhead across all applications is just 0.19%. For RAM overhead, it mainly consists of the allowlist stored in the Secure World, costing less than 80 bytes since the biggest allowlist contains less than twenty legal integer IDs.

5.2   Effectiveness Analysis

Attack detection via *CEFI*. Although privilege separation vulnerabilities are common (recent ones include for example CVE-2020-26072, CVE-2021-33220, and CVE-2022-36782), unfortunately firmware is rarely available open source. As such, there is no known vulnerable firmware available for us to test. Instead, we injected the vulnerability shown in Fig. 2 into the test example, and verified that *CEFI* could prevent attacks that exploit this vulnerability while running the program. In order to construct the vulnerability, we added some functions to the test example, such as switch_on, update_firmware, change_passwd, and so on. For the experiment, we formulated the following rules: some commands can only be issued by the cloud, and the others only by local clients. The details are shown in Table 3. However, we do not implement strict authentication, resulting in the vulnerability that the remote and local command sets can be mixed together.

In order to realize our enhancement scheme, we carried out the following steps: first, we annotate the command execution functions and entry functions, such as *switch_off*, *update_firmware*, etc. Then, the legal IDs of the paths between entry function and command execution function can be statically obtained. We traverse all legal operations, and store all legal ID information in TrustZone to form an allowlist. After getting the allowlist, we can send instructions to the device at will: for example, send the *update_firmware* command from the IoT cloud to the device, and the device can normally perform the corresponding operation, or send the *update_firmware* command from the local client to the device, and the device prompts that the operation cannot be performed. It can be seen from this that our scheme is effective. Although this vulnerability is constructed by ourselves, it tests the most important logic of the vulnerability, and our solution can indeed discover and defend against such vulnerabilities during program operation.

Security Analysis. Our threat model does not assume that the attacker can hijack the original control flow and data flow, as outlined in Sect. 3.2. To evade *CEFI*, attackers need to circumvent the validation checks of *CEFI*. They would need to achieve any of the following: 1) disable the instrumentation of the calling context ID updating logic and validation checks; 2) tamper with the metadata, which includes the calling context ID and the allowlist; or 3) discover a collision where a path from the channel (entry point) to **disallowed** commands (end point) can produce an calling context ID that is included in the allowlist.

However, 1) is ruled out by the assumption that existing control flow integrity enforcement has been effectively deployed, and attacks cannot bypass the execution of instrumented trampoline functions. *CEFI* prevents 2) by storing the crucial metadata (i.e., calling context ID and allowlist) in a TrustZone-protected secure world. The metadata can only be updated by a secure API call (i.e., instrumented trampoline function call). The assumption of control flow integrity ensures that secure API call cannot be hijacked and will be correctly executed. As a result, 2) is prevented as well. Lastly, 3) is prevented by the nature of CCE algorithm [23], which guarantees the uniqueness of the ID for each specific calling context.

5.3   Annotation Effort

CEFI requires annotation (see Listing 1.3) to express relationship between the interaction channel's entry point and the end point (i.e., command execution function). Although this is a manual process, it requires only minimal effort (see Table 1). In our experiments, we anticipate that a few minutes are enough to complete annotations for a program, assuming that programmers have the knowledge of its design logic. While these programs may seem particularly small, this reflects the fact that most embedded programs are by nature required to be much smaller than regular software. We note that there are currently no standard benchmarks, but we chose these programs because they are also widely used for evaluation in related research.

6   Related Work

CEFI is the first approach to enforce integrity of command execution on embedded devices after deployment, even against attacks that violate neither static control flow nor static data flow, even though such vulnerabilities are common on embedded devices [35]. The only other work that can identify such vulnerabilities, Gerbil [30], uses symbolic execution to find them in the testing phase, but cannot prevent exploitation of residual vulnerabilities after deployment. *CEFI*'s very low overhead makes it particularly suitable for this purpose, especially on resource-constrained embedded devices.

Other existing work focuses mostly on detecting violations of control flow and data flow. In this section, we first focus on work that enables detection of such vulnerabilities on resource-constrained embedded devices. Since context-sensitivity is critical to *CEFI*'s ability to detect privilege separation vulnerabilities, we also discuss works that introduce context sensitivity to runtime detection of violations.

6.1   Control and Data Flow Integrity on Embedded Systems

Since embedded devices are resource-constrained, solutions to enforce control-flow integrity (CFI) and data-flow integrity (DFI) are only viable if they are very lightweight. For example, μRAI [2] uses LLVM compiler passes to enforce return address integrity by removing the need to spill return addresses to the stack. Silhouette [34] leverages an incorruptible shadow stack for hardening backward indirect jump and uses a label instruction for protecting forward indirect jump. CFI CaRE [20] leverages TrustZone to implement a shadow stack mechanism. DFI is used to protect the integrity of memory access (e.g., maintaining and checking bounds information for each memory read or write). DFI requires much more instrumentation than CFI, because it needs to perform checks at memory access points rather than just at indirect branches. As such, there are fewer solutions for DFI than for CFI for embedded systems, as pointed out in a recent survey [17]. One notable work is OAT [24], selectively protects critical data on embedded programs. Therefore, it reduces performance overhead by instrumenting critical variable access, but sacrifices protection. However, this solution does not affect the protection provided by *CEFI*. Existing work cannot prevent exploitation of privilege separation vulnerabilities without violating control and data flow properties.

6.2   Context Sensitive Defense Solutions

Context sensitivity allows defenses to be more restrictive than traditional CFI and DFI solutions can be, by considering not just static properties of the control and data flow graphs, but also the actual control flow path taken at runtime. Calling context is widely used in context-sensitive defenses [9,18,19,27–29]. For example, PathArmor [27] conducts context-sensitive static analysis over the CFG on-demand, and provides context-insensitive CFI policies. However, PathArmor [27] does not protect against privilege separation vulnerabilities. Henry et al. [9] and David et al. [28] use the call stack and calling context for anomaly detection and system call trace consistency, but these methods are expensive [5]. Qiang et al. [33] propose HeapTherapy for lightweight trace collection and exploit detection, which aims to mitigate traditional heap buffer overflows. We use calling context encoding to defend against privilege separation vulnerabilities and logic bugs without control flow violations in resource-constrained embedded systems.

7   Conclusion

With the development of the Internet of Things (IoT), the application scenarios of embedded devices are becoming broader and more complicated. As a result, there are interactions between the various entities (i.e., cloud, the IoT device, mobile app). This causes the rise of a new type of vulnerability, privilege separation vulnerabilities, that can be exploited to launch attacks (e.g., device hijacking attacks) without control flow anomalies. Therefore, we propose *CEFI*-Command Execution Flow Integrity, to protect embedded devices against such attacks. Finally, we apply *CEFI* on five real-world programs. The evaluation shows that *CEFI* can effectively prevent this type of attack, with negligible runtime overhead of 0.18% and negligible memory overhead of 0.19%. In future work, we plan to evaluate *CEFI* on larger IoT systems.

Acknowledgements. We thank our shepherd Roland YAP Hock Chuan and anonymous reviewers for their valuable feedback. This work was supported by the National Natural Science Foundation of China (U1836210), the Key Research and Development Science and Technology of Hainan Province (GHYF2022010), the National Natural Science Foundation of China (No.62202188), and the National Key R&D Program of China (No.2022YFB31033400). Meanwhile, this work was partly done at VU Amsterdam. We thank the support provided by the China Scholarship Council (CSC) and the VUSec Group at VU Amsterdam.

References

1. Abera, T., et al.: C-flat: control-flow attestation for embedded systems software. In: Proceedings of the 2016 ACM SIGSAC Conference on Computer and Communications Security, pp. 743–754 (2016)
2. Almakhdhub, N.S., Clements, A.A., Bagchi, S., Payer, M.: μrai: securing embedded systems with return address integrity. In: 27th Annual Network and Distributed System Security Symposium, NDSS 2020, San Diego, California, USA, February 23–26 (2020)
3. ARM Ltd.: Arm compiler software development guide version 6.3 (2022). https://developer.arm.com/documentation/dui0773/d/chunkpge1447084556319
4. Ball, T., Larus, J.R.: Efficient path profiling. In: MICRO 29, pp. 46–57. IEEE (1996)
5. Bond, M.D., McKinley, K.S.: Probabilistic calling context. ACM SIPLAN Notices **42**(10), 97–112 (2007)
6. Cerdeira, D., Santos, N., Fonseca, P., Pinto, S.: SoK: understanding the prevailing security vulnerabilities in trustzone-assisted tee systems. In: 2020 IEEE Symposium on Security and Privacy (SP), pp. 1416–1432. IEEE (2020)
7. Clements, A.A., Almakhdhub, N.S., Bagchi, S., Payer, M.: Aces: Automatic compartments for embedded systems. In: USENIX Security 2018, vol. 2018, pp. 65–82 (2018)
8. Feng, B., Mera, A., Lu, L.: P2IM: scalable and hardware-independent firmware testing via automatic peripheral interface modeling. In: USENIX Security 2020, pp. 1237–1254 (2020)
9. Feng, H.H., Kolesnikov, O.M., Fogla, P., Lee, W., Gong, W.: Anomaly detection using call stack information. In: 2003 Symposium on Security and Privacy, 2003, pp. 62–75. IEEE (2003)
10. Gustafson, E., et al.: Toward the analysis of embedded firmware through automated re-hosting. In: RAID 2019, pp. 135–150 (2019)
11. Hassanshahi, B., Jia, Y., Yap, R.H.C., Saxena, P., Liang, Z.: Web-to-application injection attacks on android: characterization and detection. In: Pernul, G., Ryan, P.Y.A., Weippl, E. (eds.) ESORICS 2015. LNCS, vol. 9327, pp. 577–598. Springer, Cham (2015). https://doi.org/10.1007/978-3-319-24177-7_29

12. Hassanshahi, B., Yap, R.H.C.: Android database attacks revisited. In: AsiaCCS 2017, pp. 625–639 (2017)
13. Hu, H., Shinde, S., Adrian, S., Chua, Z.L., Saxena, P., Liang, Z.: Data-oriented programming: on the expressiveness of non-control data attacks. In: 2016 IEEE Symposium on Security and Privacy (SP), pp. 969–986. IEEE (2016)
14. Huo, D., Cao, C., Liu, P., Wang, Y., Li, M., Xu, Z.: Commercial hypervisor-based task sandboxing mechanisms are unsecured? but we can fix it! J. Syst. Architect. **116**, 102114 (2021)
15. Jia, Y., et al.: Burglars' IoT paradise: understanding and mitigating security risks of general messaging protocols on IoT clouds. In: 2020 IEEE Symposium on Security and Privacy (SP), pp. 465–481. IEEE (2020)
16. Koeberl, P., Schulz, S., Sadeghi, A.R., Varadharajan, V.: TrustLite: a security architecture for tiny embedded devices. In: Proceedings of the Ninth European Conference on Computer Systems, pp. 1–14 (2014)
17. Mishra, T., Chantem, T., Gerdes, R.: Survey of control-flow integrity techniques for embedded and real-time embedded systems. arXiv preprint arXiv:2111.11390 (2021)
18. Newsome, J., Brumley, D., Song, D., Chamcham, J., Kovah, X.: Vulnerability-specific execution filtering for exploit prevention on commodity software. In: NDSS (2006)
19. Novark, G., Berger, E.D., Zorn, B.G.: Exterminator: automatically correcting memory errors with high probability. In: PLDI, pp. 1–11 (2007)
20. Nyman, T., Ekberg, J.-E., Davi, L., Asokan, N.: CFI CaRE: hardware-supported call and return enforcement for commercial microcontrollers. In: Dacier, M., Bailey, M., Polychronakis, M., Antonakakis, M. (eds.) RAID 2017. LNCS, vol. 10453, pp. 259–284. Springer, Cham (2017). https://doi.org/10.1007/978-3-319-66332-6_12
21. OConnor, T., Enck, W., Reaves, B.: Blinded and confused: uncovering systemic flaws in device telemetry for smart-home internet of things. In: Proceedings of the 12th Conference on Security and Privacy in Wireless and Mobile Networks, pp. 140–150 (2019)
22. Shoshitaishvili, Y., Wang, R., Hauser, C., Kruegel, C., Vigna, G.: Firmalice-automatic detection of authentication bypass vulnerabilities in binary firmware. In: NDSS, vol. 1, p. 1 (2015)
23. Sumner, W.N., Zheng, Y., Weeratunge, D., Zhang, X.: Precise calling context encoding. IEEE Trans. Software Eng. **38**(5), 1160–1177 (2011)
24. Sun, Z., Feng, B., Lu, L., Jha, S.: OAT: attesting operation integrity of embedded devices. In: SP, pp. 1433–1449. IEEE (2020)
25. Szekeres, L., Payer, M., Wei, T., Song, D.: SoK: eternal war in memory. In: 2013 IEEE Symposium on Security and Privacy, pp. 48–62. IEEE (2013)
26. Tian, Y., et al.: Smartauth: User-centered authorization for the internet of things. In: USENIX Security 2017, pp. 2–8 (2017)
27. Van der Veen, V., et al.: Practical context-sensitive CFI. In: CCS, pp. 927–940 (2015)
28. Wagner, D., Dean, R.: Intrusion detection via static analysis. In: Proceedings 2001 IEEE Symposium on Security and Privacy. S&P 2001, pp. 156–168. IEEE (2000)
29. Xu, J., Ning, P., Kil, C., Zhai, Y., Bookholt, C.: Automatic diagnosis and response to memory corruption vulnerabilities. In: CCS, pp. 223–234 (2005)
30. Yao, Y., Zhou, W., Jia, Y., Zhu, L., Liu, P., Zhang, Y.: Identifying privilege separation vulnerabilities in IoT firmware with symbolic execution. In: Sako, K., Schneider, S., Ryan, P.Y.A. (eds.) ESORICS 2019. LNCS, vol. 11735, pp. 638–657. Springer, Cham (2019). https://doi.org/10.1007/978-3-030-29959-0_31

31. Yuan, B., et al.: Shattered chain of trust: understanding security risks in cross-cloud iot access delegation. In: USENIX Security 2020, pp. 1183–1200 (2020)

32. Zeng, Q., Rhee, J., Zhang, H., Arora, N., Jiang, G., Liu, P.: DeltaPath: precise and scalable calling context encoding. In: Proceedings of Annual IEEE/ACM International Symposium on Code Generation and Optimization, pp. 109–119 (2014)

33. Zeng, Q., Zhao, M., Liu, P.: Heaptherapy: an efficient end-to-end solution against heap buffer overflows. In: DSN 2015, pp. 485–496. IEEE (2015)

34. Zhou, J., Du, Y., Shen, Z., Ma, L., Criswell, J., Walls, R.J.: Silhouette: efficient protected shadow stacks for embedded systems. In: USENIX, pp. 1219–1236 (2020)

35. Zhou, W., et al.: Reviewing IoT security via logic bugs in IoT platforms and systems. IEEE Internet Things J. **8**(14), 11621–11639 (2021)

36. Zhou, W., et al.: Discovering and understanding the security hazards in the interactions between IoT devices, mobile apps, and clouds on smart home platforms. In: USENIX, pp. 1133–1150 (2019)

UNTANGLE: Aiding Global Function Pointer Hijacking for Post-CET Binary Exploitation

Alessandro Bertani[✉], Marco Bonelli, Lorenzo Binosi, Michele Carminati,
Stefano Zanero, and Mario Polino

Politecnico di Milano, Milan, Italy
{alessandro.bertani,lorenzo.binosi,michele.carminati,stefano.zanero,
mario.polino}@polimi.it, marco.bonelli@mail.polimi.it

Abstract. In this paper, we combine static code analysis and symbolic execution to bypass Intel's Control-Flow Enforcement Technology (CET) by exploiting function pointer hijacking. We present UNTANGLE, an open-source tool that implements and automates the discovery of global function pointers in exported library functions and their call sites. Then, it determines the constraints that need to be satisfied to reach those pointers. Our approach manages naive built-in types and complex parameters like structure pointers. We demonstrate the effectiveness of UNTANGLE on 8 of the most used open source C libraries, identifying 57 unique global function pointers, reachable through 1488 different exported functions. UNTANGLE can find and verify the correctness of the constraints for 484 global function pointer calls, which can be used as attack vectors for control-flow hijacking. Finally, we discuss current and future defense mechanisms against control-flow hijacking using global function pointers.

Keywords: Binary Exploitation · Control-Flow Integrity · Control-Flow Hijacking · Static Analysis · Symbolic Execution

1 Introduction

Binary exploitation is a significant problem and threat due to memory corruption vulnerabilities [36] in programs written using memory-unsafe languages like C. Despite this flaw, C is still widely used for its reliability, portability, and performance. Most memory corruption exploits aim to disrupt a program's control flow. Recent defense proposals primarily focus on preserving the control flow, to prevent memory corruption vulnerabilities from being exploited to redirect it on an unintended path. The main idea behind control-flow preservation is to perform checks to ensure that only allowed execution paths are taken so that any deviation from them would be recognized as malicious and stopped. An example of a state-of-the-art control-flow hijacking defense mechanism is Intel's **Control-Flow Enforcement Technology**(CET) [33], which was designed to protect both forward edges (function calls and jumps) and back edges (function returns) in the **Control-Flow Graph** (CFG) of a program. Defense mechanisms like CET make it significantly harder for an attacker to gain arbitrary code execution, as they drastically reduce the possible attack surface. With such defense mechanisms in place, an attacker cannot directly tamper with the return address of a function but must target other control variables like **function pointers**, which can be found in different memory sections of a program or library and constitute a possible attack surface.

This work focuses on **global** function pointers defined in C libraries. Global function pointers are easy to identify and find in process memory, and finding such an attack vector in a widespread library makes the approach generic, enabling the exploitation of binaries having the library as a dependency. Finding global function pointers in C libraries would simplify exploit writing in CET-enabled scenarios and would be helpful to C library developers to detect the presence of such attack vectors. To better understand how an attacker can exploit function pointers, consider a library that exports a function containing a call to a global function pointer defined within the library. This exported function is then used by a program using the library. If this program presents an arbitrary write vulnerability (i.e., a vulnerability that allows the attacker to write any value to any memory location), it can be used to overwrite the global function pointer and redirect the control flow of the program once a call to it is reached through the exported library function. There are a few complications to this kind of attack. First of all, global function pointers must be found inside a library. Even if the source code of the target library is available, one would need to manually analyze it to find all possible global function pointers, interesting call sites, and all the conditions leading the execution to them. Doing all this by hand, potentially for several different libraries, is a feasible but highly time-consuming and demanding task. Our work proposes an approach based on static analysis and symbolic execution [25] to automate this whole process given the source code of a C library. Moreover, we identify and solve all the constraints that need to be satisfied to reach global function pointer calls at runtime.

We present UNTANGLE[1], an open-source tool that implements the proposed approach to aid binary exploitation through global function pointer hijacking. It is important to highlight that UNTANGLE is also helpful from a defense perspective since it helps C library developers to discover the identified attack surface and library users to detect affected libraries. UNTANGLE performs its task through four main components: the *Global Pointers Extractor*, the *Instrumenter*, the *Parser* and the *Executor*. The *Global Pointers Extractor* performs source code analysis on the target library to find global function pointers and their call sites. The *Instrumenter* instruments the source code of the target library to prepare it for symbolic execution. The *Parser* extracts information on structure types definitions and function signatures to improve the symbolic execution process. The *Executor* performs symbolic execution on the instrumented library binaries and employs a custom memory model designed to ease handling complex

---

[1] https://github.com/untangle-tool/untangle.

function arguments. We evaluate UNTANGLE on several open-source C libraries (i.e., `libgnutls`, `libasound`, `libxml2`, `libfuse`, `libcurl`, `libnss`, `libpcre` and `libbsd`). UNTANGLE identifies 64 unique global function pointers (57 of which are reachable through exported functions) and 1488 exported functions that lead to their calls, finding and verifying the constraints' correctness to satisfy those calls in 484 cases. In summary, the contributions are the following:

- A methodology to identify global function pointers, their calls sites reachable through exported library functions, and how to reach them.
- UNTANGLE, an automatic tool that implements this methodology end-to-end. It takes the source code of a library as input and produces as output all the function pointers found inside the library, which are reachable through exported functions and concrete parameter values that satisfy the conditions that allow it to reach them.
- An ad-hoc symbolic execution memory model (implemented in UNTANGLE) that deals with `struct` pointers passed as function parameters.

2   Background

Static Code Analysis. Static code analysis is the practice of analyzing a program without executing it, and is a widely adopted technique for vulnerability research. It can be performed at the source code level (given the source code of a program or a library) or at the binary level (given the compiled program or library). In our work, we perform static source code analysis to identify global function pointers in library code. To perform this task, we use CodeQL [2], a static analysis framework developed by GitHub, that provides a formal query language to specify the targets of the static analysis process.

Symbolic Execution with `angr`. Symbolic execution is a dynamic program analysis technique in which the program to be analyzed is driven through its execution by a specialized interpreter, known as *symbolic execution engine*. The engine feeds the program with *symbolic inputs*, rather than concrete inputs obtained by the user or the environment. Whenever the analyzed program needs to evaluate a branch condition involving symbolic data, the engine creates two expressions constraining the symbolic data: one that satisfies the condition and one that does not. Then, it duplicates the current state of the program, and two initially identical states are advanced in parallel on the two different sides of the branch, keeping track of the constraints on symbolic variables that caused the state duplication. A critical aspect of a symbolic execution engine is its *symbolic memory model*, which defines the policies for managing memory accesses. Because of its Python-based interface, flexibility, and modular plugin system, we chose `angr` [1,35,37] as a symbolic execution engine. `angr`'s memory model, already analyzed in previous works [12], is fully symbolic, i.e., it emulates every memory operation by concretizing memory addresses whenever it is needed.

When dealing with a symbolic address, at first `angr` evaluates how large the range of values it can assume is. In the case of a single possible value (depending on the constraints present in the current state), the address is concretized and the load/store is performed at the concrete address. However, in the case of multiple possible values, the behavior differs between load and store operations. For a *store* operation, a symbolic address is always concretized to the maximum possible value satisfying its constraints. This can be useful if the objective of symbolic execution is to find memory corruption bugs in the analyzed program. For instance, if an unconstrained 64-bit symbolic pointer is dereferenced for a store of size 8, its value could be concretized to `0xfffffffffffffff8`. For a *load* operation, if the range of possible values exceeds a fixed internal threshold, the symbolic address is concretized to an arbitrary value returned by the solver. Otherwise, if the range is small enough, an If-Then-Else expression is generated and the address remains symbolic. The issue with the first case is that unpredictable concrete addresses could be generated, which will likely be colliding with the addresses of other existing objects. These issues can impact the chance of successfully traversing the call chain needed to reach the calls to function pointers we are interested in during symbolic execution. Complex data types, such as pointers to structure, are an especially problematic case: **angr** has no knowledge about struct sizes, and this can cause instances where addresses of different struct pointers are concretized to contiguous values. This is likely to cause memory overlaps and generate invalid results.

2.1   Exploitation Techniques and Defenses

Return-Oriented Programming (ROP) [13, 16, 28, 29]. It is a code-reuse technique that allows the execution of an arbitrary sequence of instructions in a program without injecting any code. This technique uses a "ROP-chain": a chain of short sequences of instructions, called "gadgets", that end with a *return* instruction (thus the name of the technique). ROP gadgets can be found in the code section of the target binary or any shared library loaded by it and thus visible in its address space. By chaining multiple gadgets together, each executing one or more instructions before returning, an attacker can create an arbitrary sequence of machine instructions. Given the right gadgets, ROP is also Turing-complete [21] and can execute arbitrary code. The only limits to this technique are the length of the initial ROP-chain, limited by the number of bytes that can be written on the stack past the saved return address, and the gadgets available for use, which depend on the specific program and the libraries it uses. ROP defeats defense mechanisms such as **Write Xor Execute**($W \oplus X$) since all the gadgets involved in the ROP-chain are located in executable memory pages. Code reuse techniques also include **Jump-Oriented Programming** (JOP) [11] and **Call-Oriented Programming** (COP) [30]. JOP is a code reuse technique that builds and chains gadgets that end with an *indirect branch* instruction rather than a return instruction. This eliminates reliance on the stack and return-like instructions (e.g., a stack *pop* followed by a *jump* to the popped value). COP is a similar code reuse technique that uses gadgets that end in a *call* instruction.

Defense mechanisms directly affect the impact of code-reuse techniques. **Address Space Layout Randomization** (ASLR) [9, 32] randomly arranges the address space of a process before starting its execution: the base address of different memory regions (such as the program itself, library code, stack, and heap) changes with every new execution of the same program. ASLR can andomize the position of a program in memory only if the program is a **Position-Independent Executable** (PIE), that is, a program that can properly run regardless of its position in memory. All the memory accesses of a PIE are defined using relative offsets rather than absolute addresses so that the base address where the program is loaded in memory can be arbitrarily chosen and randomly generated to be different for each execution. This mechanism strongly impacts the previously discussed exploitation techniques: a ROP-chain cannot be built without knowing the exact address of each gadget. ASLR is, however, only effective as long as a potential attacker cannot *leak* the address of an interesting memory area (e.g., a section of the program binary itself, a loaded library). If the exact address of any piece of code and data contained within it can be leaked through vulnerabilities of a program, an attacker would then be able to compute the exact address of any piece of data contained within it, as offsets inside the binary are fixed. Recent defense solutions are directly targeted at defeating ROP: some examples are kBouncer [27], ROPdefender [20] and ROPecker [18]. While targeted towards ROP, however, neither of these solutions can detect and defeat other code-reuse attacks. It has been shown in previous works that these defenses have some shortcomings and can be bypassed with low effort [15,31]. Other proposals target the preservation of the control flow of a program rather than the mitigation of a specific exploitation technique. **Control-Flow Integrity** (CFI) [8,10,14,26] is a security policy dictating that software execution must only follow paths of its CFG, which is determined ahead of time through source-code analysis, binary analysis or execution profiling. CFI paved the way for a series of defenses against control-flow hijacking attacks in hardware and software solutions.

Intel's **Control-Flow Enforcement Technology** (CET) is one of the most recent and advanced CFI enforcement defenses, providing a CPU instruction set architecture extension that allows the software to easily set up hardware defenses against ROP, JOP and COP style attacks. CET has two main features: ① the use of a **Shadow Stack** [19] to provide saved return address protection, preventing ROP; ② **Indirect Branch Tracking** (IBT) [27] to prevent the misuse of indirect branch instructions, typical of JOP/COP attacks. CET is available on all Intel Core CPUs starting from the 11th generation, and AMD recently announced CET support from its "Series 5000" processors onward. However, operating systems' support towards CET is still partial. Because of its accuracy in protecting both forward and back edges in a CFG, full-CET support in both kernel and user space would make code reuse techniques relying on overwriting the saved return address on the stack (ROP) impossible, and the ones relying on indirect control transfer instructions (JOP, COP) harder, as control-flow would need to be redirected to legitimate targets, identified by *endbranch* instructions.

Function Pointer Hijacking. If an attacker wants to redirect the control flow of a program but cannot tamper with the saved return address on the stack because there are protection mechanisms such as CET in place, they must target other kinds of control data, such as **function pointers**. Common reasons for function pointer usage in C library code are providing the user with runtime *hooks* for particular function invocations, implementing function callbacks, and delivering notifications for asynchronous runtime events. To gain arbitrary code execution through a function pointer in a CET-enabled environment, an attacker needs to consider two main possible scenarios. In the first scenario, when only the shadow stack is active, the attacker can overwrite the function pointer with any address pointing to a memory section containing executable code. In the second scenario, when IBT is active (regardless of shadow stack usage), the attacker necessarily needs to overwrite the function pointer with the address of an *endbranch* instruction. This could be the start of an interesting function, a `case` of a `switch` statement compiled using a jump table, or similar. In case the target is a function, the ability to control the parameters supplied to the function could also be necessary (e.g., targeting the `system()` function provided by the standard C library, one would need to pass the command to run as a parameter), and depends on the specific case at hand. We focus on hijacking global function pointers in C libraries as a possible exploitation entry point, considering that given the right conditions, this technique can be used to circumvent Intel CET.

3   Related Work

Non-control data attacks [17, 22] are state-of-the-art binary exploitation techniques, and are a viable alternative to "traditional" control-flow hijacking attacks. They aim at redirecting the program's control flow without tampering with control data, acting only on non-control data, such as variables used by the program to make control decisions. Data-oriented attacks are thus capable of changing the control flow of a program by bypassing defense mechanisms that preserve control-flow integrity. Sophisticated non-control data attack techniques and tools that help automate exploitation have been proposed in recent years. **Data-Oriented Programming** (DOP) [23] is a technique to construct expressive non-control data exploits. It allows an attacker to perform arbitrary computations in program memory by chaining the execution of short sequences of instructions, called DOP gadgets. It is a powerful technique, with the downside that the gadget chains must be crafted by hand. **Block-Oriented Programming** (BOP) [24] is a further improvement of data-oriented attacks: it uses basic blocks as gadgets and leverages symbolic execution to automatically find the constraints on variables and memory-resident data needed to redirect the control flow. BOP attacks are specifically aimed at creating a chain of basic blocks that does not trigger CFI preservation mechanisms, and since they do not overwrite the saved return address, they can bypass shadow stacks too. The advantage of BOP, with respect to DOP, is that the gadget chain-building process is automated. These techniques, however, have their limitations: they are complex and only work in particular situations. Global function pointer hijacking requires less effort and is a viable alternative to perform binary exploitation in specific settings.

Discussion. To the best of our knowledge, no existing work explores the automation of both global function pointer identification and hijacking in library code. Most of the existing work and research focuses on subsequent exploitation steps instead. In particular, the BOP Compiler (BOPC) [24], could benefit from our work: one of the requirements for the tool to work correctly is an entry point, i.e., a point from which the tool starts its analysis and constructs the basic block chain. A function pointer that can be overwritten with an arbitrary address would be a good starting point for this analysis.

4   Threat Model and Problem Statement

Our exploitation scenario considers a program running on a machine employing state-of-the-art control-flow hijacking defenses, such as fully enabled Intel CET. Moreover, the program is also protected through stack canaries, $W \oplus X$ memory protection policies, and ASLR. We assume that the program uses functions exported by a C library (statically linked or dynamically loaded at runtime) that contain, or can lead to, calls to global function pointers defined within the library itself. We assume that the program presents a known memory corruption vulnerability that can lead to an arbitrary memory write, also known as "write-what-where" primitive, which gives an attacker the ability to write any value to any writable address. In the case of a dynamic library, we also assume that the attacker can discover, for example, thanks to an information leak, the base address at which the target was loaded under ASLR. These assumptions are realistic and practical since they are in line with the ones of the mechanisms that aim at preventing arbitrary memory reads and writes from being exploited.

Motivation and Research Goal. One of the fundamental steps while writing an exploit that aims at gaining arbitrary code execution is to gain control of the instruction pointer. This is usually achieved by overwriting the saved return address of a function on the stack, by overwriting a function pointer contained in an object on the stack or on the heap (e.g., a *vtable* pointer), or by overwriting global function pointers (e.g., in shared libraries). If CFI enforcement mechanisms like Intel CET are in place, the first approach cannot be applied because of the shadow stack. The second approach strongly depends on the specific application the attacker wants to exploit, while the last approach is more general and can be applied to any application using the same shared libraries. Being able to find global function pointers in libraries would simplify exploit writing for such applications. For this reason, our goal is to find ① global function pointers calls in the source code of a target library; ② the conditions to reach such calls, giving us the ability to gain arbitrary code execution. Commonly used C libraries can be composed of hundreds or even thousands of source code files, while the total number of lines of code can vary from a few thousand to several hundred thousand. Manually searching for global function pointers and all locations where they are called is feasible but not trivial: analyzing a large code base would require considerable time and effort. Even if we could find all function pointers and calls manually, it is challenging to identify the conditions over function parameters and other global variables that would lead the program to the execution of such calls. In fact, some libraries contain functions that are hundreds of source code lines long. Manually keeping track of all the conditions needed to be satisfied to reach a specific code section at runtime would be demanding, time-consuming, and error-prone. Therefore, automating the identification of global function pointer calls is necessary. This would make the whole process faster, more practical, and more reliable, and would provide library developers with an effective way to identify and reduce the attack surface in their code.

5   UNTANGLE

UNTANGLE uses a combination of static analysis, library source code instrumentation, and symbolic execution to provide precise information on how to reach global function pointer calls starting from exported functions of a given C library. This includes information on the constraints on function parameters and global variables that need to be satisfied to reach these calls. The workflow of UNTANGLE includes several components: the *Global Pointers Extractor*, the *Instrumenter*, the *Parser* and the *Executor*, which contains a custom memory model for symbolic execution. The *Global Pointers Extractor* creates a CodeQL database for the library from its source. CodeQL's query language allows specifying precisely the targets of the static analysis: in our case, the targets are global function pointers, their call sites, and library functions that can reach them, along with their signatures. After the creation of the database, the *Global Pointers Extractor* performs queries to identify these targets. The *Instrumenter* then places a call to a uniquely generated *target* function immediately before each identified global function pointer call, and builds a new, instrumented version of the library. The *Parser* performs two different tasks: *struct* parsing and *function signature* parsing. The results of both these tasks are passed to the *Executor* component: the information on function signatures is used by the symbolic execution engine, while the information on structures is used by the custom memory model. The *Executor* uses the instrumented library binaries to evaluate the reachability of identified global function pointer calls, treating the functions inserted by the *Instrumenter* as targets to reach. The actual symbolic execution

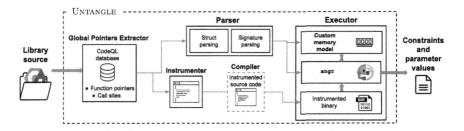

Fig. 1. Architecture overview of UNTANGLE.

is performed by angr. We use angr because its modular design allows us to easily add new functionality or modify existing behavior. In fact, the *Executor* uses a *Symbolic Memory Model* for angr, developed with its plugin system, specifically designed to ease the handling of complex structure pointer parameters. The *Executor* also has a built-in *Automatic Result Validation Mechanism*, that we use to test the correctness of the results of the symbolic execution phase. For an overview of the architecture of our tool, refer to Fig. 1.

Global Pointers Extractor. This component performs the static analysis of the source code of the library, which is provided as an input to UNTANGLE. As previously mentioned, we use CodeQL for the static analysis, as it allows us to accurately specify the targets of the analysis through its formal query language.

First, the *Global Pointers Extractor* builds a CodeQL database along with the original library. Then, it runs two queries[2] on the database. The first query performs three simultaneous operations: ① detection of all existing global function pointer variables; ② identification of all the call sites for each detected variable; ③ discovery of potential entry points to reach the call sites. The last operation involves traversing CodeQL's call graph, starting from any function containing one or more call sites, going backward from callee to caller, and listing all non-static library functions encountered. We can check whether an identified library function is exported by looking at the exported symbols of the library binaries. The second query detects structure definitions, the fields they are composed of, and their offsets inside the structure. The results of this query are passed to the *Parser*, which will use them to create and manage internal objects representing structure pointers.

Instrumenter. The purpose of the *Instrumenter* is to provide targets for the symbolic execution phase through source code instrumentation. This phase must preserve the original functionality of the library to allow the symbolic execution phase to provide reliable results. For this purpose, the *Instrumenter* inserts a call to a uniquely named dummy target function right before each global function pointer call found by the *Global Pointers Extractor*. This new call has only one artificial side effect that prevents it from being optimized away by the compiler. The instrumented library source code is then re-compiled, and the resulting binaries contain exported symbols referencing the newly inserted target functions. This allows providing angr with precise indications on the target addresses.

Parser and Executor. UNTANGLE can find constraints on parameters of exported functions and global variables that need to be satisfied to reach identified global function pointer call sites and then evaluate them to find suitable concrete values. The *Parser* extracts the number and types of parameters from the signature of each function that needs to be symbolically executed, creating symbolic bit-vectors of the appropriate size. For struct pointer parameters, the *Parser* also creates the needed StructPointer objects as previously discussed. The *Symbolic Memory Model* uses these objects to handle symbolic memory

---

[2] https://github.com/untangle-tool/untangle/blob/main/untangle/analyzer.py# L82.

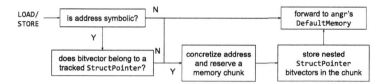

Fig. 2. Load/Store handling using UNTANGLE's memory model

loads and stores to structure pointers during the symbolic execution. To allow the identification and evaluation of interesting global variables, the *Executor* transforms writable data sections of the library binary (`.bss` and `.data`) to symbolic bit-vectors to verify later whether any memory regions belonging to these sections were involved in any constraints. This also allows the detection of constraints on the global function pointers themselves. However, these constraints depend on the specific library being tested and need to be evaluated case by case.

Symbolic Memory Model. Angr's default address concretization strategy can cause memory overlaps since it is unaware of variable types and sizes. Therefore, it cannot reserve specific regions of memory for symbolic pointers. A prime example is pointers to `struct` types. To correctly handle `struct` pointers, UNTANGLE extends angr's memory model implementing ad-hoc logic. This logic is summarized in Fig. 2. Function arguments that are pointers to known `struct` types, extracted through CodeQL, are recursively parsed into an internal `StructPointer` object, which holds fields' offsets, sizes, and symbolic bit-vectors. During symbolic execution, UNTANGLE keeps track of `StructPointer` objects to handle load/store memory operations involving their addresses. The first load/store operation through the symbolic bit-vector of a tracked `StructPointer` p concretizes its value to an address determined by a simple bump allocator. At this address, UNTANGLE reserves a chunk of symbolic memory of the needed size to hold the content of the underlying `struct` that p is tracking. Then, UNTANGLE stores the symbolic bit-vectors for any nested `StructPointer` field of p at the correct offset in the chunk. Any subsequent load/store operation to the now-concrete address is then forwarded to angr's default handler. Using this approach recursively, UNTANGLE can also handle *nested* struct pointers.

Automatic Result Validation Mechanism. UNTANGLE is equipped with an automatic result validation mechanism. Validation is performed by compiling and running a test C program that uses the solution found through symbolic execution to appropriately set up a function call to the tested library function. This is not a simple task, and depending on the library, the test program would need to be significantly complex to compile correctly. Using a library function means importing the correct header files, creating variables of the appropriate type and value (which can, in turn, require additional headers for the type definitions), and linking the right library binary after compilation. Doing this requires

Table 1. List of tested libraries and number of global function pointers found in each library by UNTANGLE.

Library	Estimated lines of source code	Unique global function pointers	Reachable function pointers
libgnutls v3.6.16	422 804	15	14
libasound v1.2.4	94 288	3	2
libxml2 v2.9.10	353 481	8	6
libfuse v3.11	21 568	1	1
libcurl v7.84	152 921	5	5
libnss v2.31	10 568	21	18
libpcre v8.39	107 530	3	3
libbsd v0.11.3	11 316	8	8
Total	1 174 476	64	57

multiple steps that change based on the specific library and cannot be easily done programmatically. We have implemented a simpler automatic verification method that involves the use of `libdl` [5] to dynamically load instrumented libraries at runtime and **The GNU Debugger** (GDB) to monitor whether identified call sites are reached through automatically inserted breakpoints. The goal of this built-in *Automatic Validation Mechanism* is to avoid false positive results: if execution reaches a breakpoint set at the target call site while running under GDB, the solution must inevitably be correct (it could be trivial, but correct). Therefore, an incorrect solution will never pass validation. This mechanism can, however, yield false negatives: functions for which UNTANGLE found a solution but through which the global function pointer call site is not reached during automatic validation. These are more complex to handle and require manual testing to be identified. Automatic validation consists of the following steps performed after a symbolic execution run that found a satisfiable solution: ① Generate and compile a C program that loads the tested library using `libdl` and calls the target function using parameter values taken from the solution; ② Run the compiled program under GDB, setting a breakpoint on the target function corresponding to the global function pointer call site that needs to be reached; ③ Check whether the breakpoint is reached or not.

6 Experimental Validation

In order to test UNTANGLE we performed *full library execution* tests on multiple C libraries commonly used on GNU/Linux systems. The main focus of our tests was the symbolic execution phase: the success rate of symbolic execution (i.e., what percentage of runs can find and return a solution), the validity of found solutions, and the number of system resources needed to find them. We collected statistics about the quantity and validity of symbolic execution results, then about performance in terms of execution time and memory usage.

Table 2. Number of unique call sites, exported functions, and unique paths to global function pointer calls for each tested library.

Library	Unique call sites	**Exported functions**	Unique paths to call sites
libgnutls v3.6.16	1 338	827	29 817
libasound v1.2.4	383	243	7 739
libxml2 v2.9.10	2 125	225	254 096
libfuse v3.11	110	110	110
libcurl v7.84	271	48	11 238
libnss v2.31	34	15	74
libpcre v8.39	13	12	36
libbsd v0.11.3	8	8	8
Total	4 282	1 488	303 118

Dataset. We selected top-ranked free, open-source C libraries listed under the "libs" section of the **Debian package Popularity Contest** [3], using the latest version provided by Debian 11 packages. We checked the presence of global function pointers using CodeQL and analyzed them with UNTANGLE. We manually compiled and checked around 50 libraries, found 8 of them (listed in Table 1), to contain interesting function pointers, and we tested them.

Experimental Setup. As shown in Table 2, the number of unique code paths starting from exported library functions and leading to a global function pointer call can be quite large. Hence, we did not test every single path, as the amount of time needed for such kind of analysis would have been prohibitive, but rather focused on analyzing the reachability of *any* global function pointer call starting from every single exported function. The machine used for testing is equipped with a 64-bit Intel Core i9-10900 CPU (base core clock speed of 2.80 GHz), 32 GiB of RAM, and runs Debian 11 GNU/Linux v5.10. Libraries were therefore compiled for Linux x86-64 using The **GNU C Compiler** (GCC) version 10.2.1, the standard compiler for Debian GNU/Linux systems. Where possible and permitted by library configuration scripts, the optimization option chosen was -O2, and the use of advanced CPU-specific instruction sets (e.g., AVX2, SSE4) was disabled to avoid issues with PyVEX [7,34], the Python library used by **angr** for translation of machine instructions. Since **angr** does not offer multi-threading support, all performed symbolic execution runs consist of single-threaded processes. Each symbolic execution run was limited to 15 min and 16GiB of RAM usage (Resident Set Size). Runs exceeding any of the two limits were halted while still collecting resource usage information for statistical purposes.

6.1 Symbolic Execution Results

The static analysis results found by UNTANGLE are listed in Table 1 and Table 2. The first table shows the number of *unique* global function pointers found in each

Table 3. Symbolic execution results and validation of successful runs.

Library	Tested functions	Symbolic execution solution		Validation result	
		Found	Not Found	Pass	Fail
libgnutls	827	460 (55.6%)	367 (44.4%)	272 (32.9%)	188 (22.7%)
libasound	243	153 (63.0%)	90 (37.0%)	91 (37.4%)	62 (25.5%)
libxml2	225	139 (61.8%)	86 (38.2%)	60 (26.7%)	79 (35.1%)
libfuse	110	59 (53.6%)	51 (46.4%)	15 (13.6%)	44 (40.0%)
libcurl	48	40 (83.3%)	8 (16.7%)	30 (62.5%)	10 (20.8%)
libnss	15	9 (60.0%)	6 (40.0%)	2 (13.3%)	7 (46.7%)
libpcre	12	9 (75.0%)	3 (25.0%)	6 (50.0%)	3 (25.0%)
libbsd	8	8 (100%)	0	8 (100%)	0
Overall	1488	877 (58.9%)	611 (41.1%)	484 (32.5%)	393 (26.4%)

library: we ruled out the ones that were not *reachable* through manual analysis. The output of the static analysis contains a list of all global function pointers identified and every library function that can reach a call to one of them. Table 2 presents the number of *exported* functions able to reach a global function pointer call. Functions that are not exported cannot be called from a program that uses the library, so they are not interesting for our tests: while testing, we check in the compiled library binary if a function is exported or not and perform symbolic execution only on exported functions. As shown in Table 3, we found a solution for 58.9% (877) of the 1488 total exported library functions analyzed.

As explained in Sect. 5, UNTANGLE has a built-in validation mechanism, which is necessary to understand which of the solutions found through symbolic execution are correct. Validation results are also summarized in Table 3: out of the 877 solutions found, 484 of those (55.2% of the found solutions, 32.5% of the total tests) were proven to be valid using the *Automatic Validation Method* described before. We can also notice the result of what we explained in Sect. 2: instances, where pointers to primitive types need to be passed as function arguments, can be concretized by **angr** to invalid memory addresses, which can make automatic validation fail. Due to this reason, even if UNTANGLE was able to find a solution that did not pass validation, there is a chance that such an instance is a false negative. UNTANGLE will report the solution, but manual testing is needed to understand additional and possibly more complex constraints that were not automatically identified. Finally, looking at runs that did not result in a found solution, we can break down the reason into four categories (shown in Table 4): *Unreachable, Timeout, Memory, Engine Error.*

Unreachable refers to a completed symbolic execution, but the engine determined that none of the identified call sites is reachable. Apart from **angr**'s limitations we discussed in Sect. 2, this can happen because the constraints leading to call sites are impossible to satisfy. *Timeout* refers to a run halted after exceeding 15 min. *Memory* refers to a run halted after exceeding 16GiB of used memory.

Engine error refers to a run halted because of an internal error of the symbolic execution engine. This happens for multiple reasons, the most common of which are constraints that become too complex (e.g., causing the solver to exceed Python's maximum call stack size) or bugs in the engine code. As we can see from Table 4, the first category is the least common. The most common failure reason is running out of memory. 16GiB is a reasonable amount of RAM; exceeding it indicates accumulating too many symbolic states along the way, which ultimately results in slower running times.

Table 4. Break-down of unsuccessful symbolic execution runs

Library	Tested functions	Solution not found	Reason			
			Unreachable	Timeout	Memory	Engine error
libgnutls	827	367 (44.4%)	26 (3.1%)	24 (2.9%)	233 (28.2%)	84 (10.2%)
libasound	243	90 (37.0%)	27 (11.1%)	23 (9.5%)	11 (4.5%)	29 (11.9%)
libxml2	225	86 (38.2%)	10 (4.4%)	12 (5.3%)	41 (18.2%)	23 (10.2%)
libfuse	110	51 (46.4%)	7 (6.4%)	9 (8.2%)	1 (0.9%)	34 (30.9%)
libcurl	48	8 (16.7%)	0	3 (6.2%)	0	5 (10.4%)
libnss	15	6 (40.0%)	0	1 (6.7%)	5 (33.3%)	0
libpcre	12	3 (25.0%)	3 (25.0%)	0	0	0
libbsd	8	0	0	0	0	0
Total	1 488	611 (41.1%)	70 (4.7%)	72 (4.84%)	291 (19.6%)	175 (11.8%)

6.2   Performance Evaluation

The execution time and the memory usage for symbolic execution, as well as the overall time spent analyzing a given library, are important metrics to measure UNTANGLE's performance. As previously mentioned, we limited each symbolic execution run to 15 min and 16 GiB of RAM, and each run exceeding either one of these limits was halted. However, we still collected statistics on halted runs and included them in the computation of the results shown in Table 5. The tests we performed took 1 min and 36 s (on average) for each function that was symbolically executed, and the average memory usage was 4373 MiB. Most of the libraries we analyzed have a much lower average memory usage than the overall average memory usage. Three of the libraries (`libgnutls`, `libxml2`, and `libnss`) have a high average memory usage. This could be due to the complexity of the functions that were symbolically executed: the length of the function, the number of control decisions the function takes, and the number of other functions called inside the analyzed function are all factors that can influence the memory usage of the symbolic execution engine.

Table 5. Resource usage statistics collected by UNTANGLE.

Library	Tested functions	Runtime		Average memory usage
		Total	Average	
libgnutls	827	20 h 21 m	1 m 29 s	5 252 MiB
libasound	243	7 h 52 m	1 m 56 s	856 MiB
libxml2	225	7 h 29 m	2 m 00 s	4 889 MiB
libfuse	110	2 h 53 m	1 m 34 s	591 MiB
libcurl	48	52 m 50 s	1 m 06 s	862 MiB
libnss	15	31 m 15 s	2 m 05 s	5 746 MiB
libpcre	12	36 s	3 s	290 MiB
libbsd	8	8 s	1 s	318 MiB
Overall	1 488	40 h	1 m 36 s	4 373 MiB

7   Impact and Defenses

Impact. In the previous section, we have shown that UNTANGLE can effectively find global function pointers in library code and can also provide reliable information on how to reach a call to one of those pointers. Our work has shown how an attacker can identify global function pointers in library code, which are attack vectors even with Intel CET enabled. To highlight the relevance of the problem addressed in our work, we searched for all Ubuntu 22.04 LTS packages using the libraries we tested. The results of this search are collected in Table 6. The total number of unique packages depending on one or more of the libraries we tested is 1820. The list of all packages installed by default on Ubuntu 22.04 LTS contains 157 of these packages. This means that 8.54% of the default packages on Ubuntu 22.04 LTS (which are 1854 by default) have one or more of the libraries we tested as a dependency. A vulnerability allowing arbitrary writes in one of these packages would allow global function pointer hijacking and enable exploitation in CET-enabled scenarios.

A real-world example of such vulnerability is the heap overflow described in **CVE-2021-43527**[3] and **CVE-2021-43529**[4], affecting Network Security Services (NSS) versions prior to 3.73. This vulnerability affects email clients and PDF viewers that use NSS for signature verification, such as Mozilla Thunderbird, LibreOffice, Evolution, and Evince. NSS is one of the libraries in which we found global function pointer calls during our tests. For this reason, exploiting the heap overflow vulnerability (in any of the programs mentioned above) to perform an arbitrary memory write would enable an attacker to achieve instruction pointer control even in CET-enabled scenarios.

---

[3] https://nvd.nist.gov/vuln/detail/CVE-2021-43527.
[4] https://nvd.nist.gov/vuln/detail/CVE-2021-43529.

Table 6. Number of Ubuntu packages depending on the libraries we used in our tests. The number between parentheses is the number of unique packages (since some of them can have more than one of these libraries as a dependency).

Library	libgnutls	libasound	libxml2	libfuse	libcurl	libnss	libpcre	libbsd	Total
# of packages	252	323	699	31	180	63	209	264	2021 (1820)

Defenses. As previously mentioned, the static code analysis process implemented in UNTANGLE can help library developers to find global function pointers in their code that can be reached through exported functions. With this information, they can employ appropriate measures to prevent global function pointers from being used as attack vectors for control-flow hijacking exploits.

This paper demonstrated the relevance of securing function pointers to avoid control-flow hijacking attacks in settings where CFI defenses are in place. Indirect call protection mechanisms already exist in LLVM: *Indirect Function Call Checks* (IFCC) checks the original function pointer's signature against the signature of the function that is actually called through the function pointer. Unfortunately, this mitigation is still not adopted among the major Linux distributions as the most used among them (Ubuntu, Debian, Arch, Fedora) use GCC as the default compiler in their build systems. Consequently, until this countermeasure becomes widespread, the results of UNTANGLE can still be used for exploitation and underline the relevance of indirect call protection mechanisms. Some defense proposals are currently being developed with this goal. **FineIBT** [4] is a software defense proposal for the Linux kernel that builds over CET, adding special instrumentation to the generated binary to enforce the verification of hashes on function prologues whenever these are indirectly called. The hashes are computed over function, and *function pointer* prototypes at compile-time and checked at run-time whenever an indirect call happens.

8 Limitations and Future Work

The main limitations of UNTANGLE come from the tool used for static analysis of source code: CodeQL. As mentioned in Sect. 5, CodeQL performs its analysis at the source code level, and it does not provide any information about the location of specific instructions or basic blocks in the resulting compiled binaries. First, while it speeds up the search for global function pointer call sites in library source code with respect to manual inspection, UNTANGLE is not always able to identify *all* of the possible call sites. Instances where a call happens *indirectly* (and not through the global function pointer identifier) are not detected: for example, global function pointers might be copied into local variables, which are then used to perform the actual call later in the code, perhaps in a different function. Detecting and correctly handling such cases would require tracking variables' assignments and copies throughout the entire code base. CodeQL offers a mechanism to do this through taint analysis but would still be unable to cover all instances. An example is when the address contained in a function pointer is copied using inline assembly, which CodeQL cannot handle. Another limitation of UNTANGLE is the way instrumentation is performed. Depending on how the library is written, it is not always possible to place a function call before the identified function pointer call without changing the original semantics of the program. In fact, in specific situations where complex macros are involved, we cannot apply our instrumentation method as-is: the only way to analyze such cases is to manually expand every instance of the macro before instrumenting it (which was the case with **gnutls** in our tests). The information provided by CodeQL makes the location of global function pointer call sites only identifiable at the source code level. Extracting call site locations in the compiled library binaries would remove the need to perform instrumentation of the source code and allow for it to be performed at a later compilation stage. Frameworks like the **LLVM Compiler Infrastructure** [6] that provide introspection and instrumentation ability at the **Intermediate Representation** (IR) level or even at the machine code level could be leveraged to directly instrument the generated code. Additionally, being able to keep track of the offset within the `.text` section of the generated *call* instruction for each interesting global function pointer call site, one could provide those directly to **angr** as a target for symbolic execution. Because of its design, UNTANGLE needs the library's source code to analyze. An improvement possibility that could be explored is the extension of our approach to binaries with no source code available. Frameworks such as **Joern** [38], which enable static analysis of binary executables, could be leveraged along with heuristics to identify which call sites to consider as global function pointer calls. Searching for all the indirect calls in a binary and evaluating if they can be hijacked could be an extension of what UNTANGLE already does and could be interesting to investigate. However, this task is challenging as it would be computationally expensive to perform through symbolic execution.

9   Conclusions

This work provides an automated methodology for finding global function pointers whose calls are reachable through exported C library functions, along with all the constraints that need to be satisfied to reach them. The approach we present employs static analysis of the source code of a target library to identify global function pointer calls and interesting exported functions, combined with symbolic execution to find constraints on function parameters and global variables that need to be satisfied to reach such calls. We present UNTANGLE, a tool that implements this approach to assist manual binary exploitation through function pointer hijacking. UNTANGLE relies on an ad-hoc symbolic execution memory model that makes it possible to deal with complex objects, such as pointers to structures, passed as function parameters. The results from the tests run on UNTANGLE show that global function pointers can be found in commonly used C libraries and that, under the right conditions, it is possible to reach calls to them starting from exported library functions. Even with Intel

CET enabled, such variables offer a possibility to gain arbitrary code execution if they are overwritten with the address of a carefully chosen legitimate target. Therefore, Untangle provides a reasonable and practical exploitation aid for function pointer hijacking.

References

1. angr. https://angr.io/
2. CodeQL. https://codeql.github.com/
3. Debian popularity contest. https://popcon.debian.org/main/index.html
4. Fine-grained forward CFI on top of intel CET / IBT. https://www.openwall.com/lists/kernel-hardening/2021/02/11/1
5. Linux standard base specification: Interface definitions for libdl. https://refspecs.linuxbase.org/LSB_3.0.0/LSB-generic/LSB-generic/libdlman.html
6. The LLVM compiler infrastructure. https://llvm.org/
7. PyVEX. https://github.com/angr/pyvex
8. Abadi, M., Budiu, M., Erlingsson, Ú., Ligatti, J.: Control-flow integrity principles, implementations, and applications. ACM Trans. Inf. Syst. Secur. **13**(1), 4:1-4:40 (2009). https://doi.org/10.1145/1609956.1609960
9. Bhatkar, S., DuVarney, D.C., Sekar, R.: Address obfuscation: An efficient approach to combat a broad range of memory error exploits. In: Proceedings of the 12th USENIX Security Symposium, Washington, D.C., USA, 4–8 August 2003. USENIX Association (2003)
10. Bletsch, T.K., Jiang, X., Freeh, V.W.: Mitigating code-reuse attacks with control-flow locking. In: Twenty-Seventh Annual Computer Security Applications Conference, ACSAC 2011, Orlando, FL, USA, 5–9 December 2011, pp. 353–362. ACM (2011). https://doi.org/10.1145/2076732.2076783
11. Bletsch, T.K., Jiang, X., Freeh, V.W., Liang, Z.: Jump-oriented programming: a new class of code-reuse attack. In: Proceedings of the 6th ACM Symposium on Information, Computer and Communications Security, ASIACCS 2011, Hong Kong, China, 22–24 March 2011, pp. 30–40. ACM (2011). https://doi.org/10.1145/1966913.1966919
12. Borzacchiello, L., Coppa, E., D'Elia, D.C., Demetrescu, C.: Memory models in symbolic execution: key ideas and new thoughts. Softw. Test. Verification Reliab. **29**(8) (2019). https://doi.org/10.1002/stvr.1722
13. Buchanan, E., Roemer, R., Savage, S., Shacham, H.: Return-oriented programming: Exploitation without code injection. Black Hat 8 (2008)
14. Burow, N., Carr, S.A., Nash, J., Larsen, P., Franz, M., Brunthaler, S., Payer, M.: Control-flow integrity: Precision, security, and performance. ACM Comput. Surv. **50**(1), 16:1–16:33 (2017). https://doi.org/10.1145/3054924
15. Carlini, N., Wagner, D.A.: ROP is still dangerous: Breaking modern defenses. In: Proceedings of the 23rd USENIX Security Symposium, San Diego, CA, USA, 20–22 August 2014, pp. 385–399. USENIX Association (2014)
16. Checkoway, S., Davi, L., Dmitrienko, A., Sadeghi, A., Shacham, H., Winandy, M.: Return-oriented programming without returns. In: Proceedings of the 17th ACM Conference on Computer and Communications Security, CCS 2010, Chicago, Illinois, USA, 4–8 October 2010. pp. 559–572. ACM (2010). https://doi.org/10.1145/1866307.1866370

17. Chen, S., Xu, J., Sezer, E.C.: Non-control-data attacks are realistic threats. In: Proceedings of the 14th USENIX Security Symposium, Baltimore, MD, USA, July 31 - August 5, 2005. USENIX Association (2005)

18. Cheng, Y., Zhou, Z., Yu, M., Ding, X., Deng, R.H.: Ropecker: A generic and practical approach for defending against ROP attacks (2014)

19. Dang, T.H.Y., Maniatis, P., Wagner, D.A.: The performance cost of shadow stacks and stack canaries. In: Proceedings of the 10th ACM Symposium on Information, Computer and Communications Security, ASIA CCS 2015, Singapore, 14–17 April 2015. pp. 555–566. ACM (2015). https://doi.org/10.1145/2714576.2714635

20. Davi, L., Sadeghi, A., Winandy, M.: Ropdefender: a detection tool to defend against return-oriented programming attacks. In: Proceedings of the 6th ACM Symposium on Information, Computer and Communications Security, ASIACCS 2011, Hong Kong, China, 22–24 March 2011, pp. 40–51. ACM (2011). https://doi.org/10.1145/1966913.1966920

21. Homescu, A., Stewart, M., Larsen, P., Brunthaler, S., Franz, M.: Microgadgets: Size does matter in turing-complete return-oriented programming. In: 6th USENIX Workshop on Offensive Technologies, WOOT'12, 6–7 August 2012, Bellevue, WA, USA, Proceedings, pp. 64–76. USENIX Association (2012)

22. Hu, H., Chua, Z.L., Adrian, S., Saxena, P., Liang, Z.: Automatic generation of data-oriented exploits. In: 24th USENIX Security Symposium, USENIX Security 15, Washington, D.C., USA, 12–14 August 2015, pp. 177–192. USENIX Association (2015)

23. Hu, H., Shinde, S., Adrian, S., Chua, Z.L., Saxena, P., Liang, Z.: Data-oriented programming: On the expressiveness of non-control data attacks. In: IEEE Symposium on Security and Privacy, SP 2016, San Jose, CA, USA, 22–26 May 2016, pp. 969–986. IEEE Computer Society (2016). https://doi.org/10.1109/SP.2016.62

24. Ispoglou, K.K., AlBassam, B., Jaeger, T., Payer, M.: Block oriented programming: Automating data-only attacks. In: Proceedings of the 2018 ACM SIGSAC Conference on Computer and Communications Security, CCS 2018, Toronto, ON, Canada, 15–19 October 2018, pp. 1868–1882. ACM (2018). https://doi.org/10.1145/3243734.3243739

25. King, J.C.: Symbolic execution and program testing. Commun. ACM 19(7), 385–394 (1976). https://doi.org/10.1145/360248.360252

26. Niu, B., Tan, G.: Modular control-flow integrity. In: ACM SIGPLAN Conference on Programming Language Design and Implementation, PLDI 2014, Edinburgh, United Kingdom - 09–11 June 2014, pp. 577–587. ACM (2014). https://doi.org/10.1145/2594291.2594295

27. Pappas, V., Polychronakis, M., Keromytis, A.D.: Transparent ROP exploit mitigation using indirect branch tracing. In: Proceedings of the 22th USENIX Security Symposium, Washington, DC, USA, 14–16 August 2013, pp. 447–462. USENIX Association (2013)

28. Prandini, M., Ramilli, M.: Return-oriented programming. IEEE Secur. Priv. 10(6), 84–87 (2012). https://doi.org/10.1109/MSP.2012.152

29. Roemer, R., Buchanan, E., Shacham, H., Savage, S.: Return-oriented programming: Systems, languages, and applications. ACM Trans. Inf. Syst. Secur. 15(1), 2:1–2:34 (2012). https://doi.org/10.1145/2133375.2133377

30. Sadeghi, A.A., Niksefat, S., Rostamipour, M.: Pure-Call Oriented Programming (PCOP): chaining the gadgets using call instructions. J. Comput. Virology Hacking Techniques 14(2), 139–156 (2017). https://doi.org/10.1007/s11416-017-0299-1

31. Schuster, F., et al.: Evaluating the effectiveness of current anti-ROP defenses. In: Stavrou, A., Bos, H., Portokalidis, G. (eds.) RAID 2014. LNCS, vol. 8688, pp. 88–108. Springer, Cham (2014). https://doi.org/10.1007/978-3-319-11379-1_5

32. Shacham, H., Page, M., Pfaff, B., Goh, E., Modadugu, N., Boneh, D.: On the effectiveness of address-space randomization. In: Proceedings of the 11th ACM Conference on Computer and Communications Security, CCS 2004, Washington, DC, USA, 25–29 October 2004, pp. 298–307. ACM (2004). https://doi.org/10.1145/1030083.1030124

33. Shanbhogue, V., Gupta, D., Sahita, R.: Security analysis of processor instruction set architecture for enforcing control-flow integrity. In: Proceedings of the 8th International Workshop on Hardware and Architectural Support for Security and Privacy, HASP@ISCA 2019, 23 June 2019, pp. 8:1–8:11. ACM (2019). https://doi.org/10.1145/3337167.3337175

34. Shoshitaishvili, Y., Wang, R., Hauser, C., Kruegel, C., Vigna, G.: Firmalice - automatic detection of authentication bypass vulnerabilities in binary firmware (2015)

35. Shoshitaishvili, Y., et al.: SOK: (state of) the art of war: offensive techniques in binary analysis. In: IEEE Symposium on Security and Privacy, SP 2016, San Jose, CA, USA, 22–26 May 2016, pp. 138–157. IEEE Computer Society (2016). https://doi.org/10.1109/SP.2016.17

36. Szekeres, L., Payer, M., Wei, T., Song, D.: SOK: eternal war in memory. In: 2013 IEEE Symposium on Security and Privacy, SP 2013, Berkeley, CA, USA, 19–22 May 2013, pp. 48–62. IEEE Computer Society (2013). https://doi.org/10.1109/SP.2013.13

37. Wang, F., Shoshitaishvili, Y.: ANGR - the next generation of binary analysis. In: IEEE Cybersecurity Development, SecDev 2017, Cambridge, MA, USA, 24–26 September 2017, pp. 8–9. IEEE Computer Society (2017). https://doi.org/10.1109/SecDev.2017.14

38. Yamaguchi, F., Golde, N., Arp, D., Rieck, K.: Modeling and discovering vulnerabilities with code property graphs. In: 2014 IEEE Symposium on Security and Privacy, SP 2014, Berkeley, CA, USA, 18–21 May 2014, pp. 590–604. IEEE Computer Society (2014). https://doi.org/10.1109/SP.2014.44

Author Index

Printed in the United States
by Baker & Taylor Publisher Services